Marine Environmental Quality

To Alain

Marine Environmental Quality

Healthy Coastal Waters

Edited by

Frédéric Muttin
Hélène Thomas

iSTE

WILEY

First published 2021 in Great Britain and the United States by ISTE Ltd and John Wiley & Sons, Inc.

ISTE Ltd
27-37 St George's Road
London SW19 4EU
UK

www.iste.co.uk

John Wiley & Sons, Inc.
111 River Street
Hoboken, NJ 07030
USA

www.wiley.com

Library of Congress Control Number: 2021938329

British Library Cataloguing-in-Publication Data
A CIP record for this book is available from the British Library
ISBN 978-1-78630-758-3

FSC
www.fsc.org
MIX
Paper from responsible sources
FSC® C013604

Contents

Preface xi
Frédéric MUTTIN and Hélène THOMAS

Chapter 1. Oil Spill Boom Strategy for Local Authorities Facing Coastal and Harbor Pollutions: A Survey of the European Research Project ISDAMP+ 1
Frédéric MUTTIN and Rose CAMPBELL

1.1. Introduction. 1
1.1.1. Challenge. 2
1.1.2. Problem. 2
1.1.3. Project partners 3
1.2. Coastal hydrodynamic model 4
1.3. Boom structural analysis model. 5
1.4. Hydrodynamic and structural models coupling. 6
1.5. Full-scale experiments 7
1.5.1. Falmouth 8
1.5.2. Lisbon 10
1.5.3. Rochefort. 12
1.5.4. La Rochelle Chef de Baie 13
1.6. Conclusion 16
1.7. Acknowledgments. 17
1.8. References 17

Chapter 2. Governance and Public Policies of Coastal Water Quality in the Pertuis Charentais 21

Alice MAZEAUD, François BÉNITEZ, Anaïs RIDEL, Arno BRINGER, Christine PLUMEJEAUD, Pascal BRUNELLO, Adélaïde LEREBOURS, Yann RABUTEAU and Hélène THOMAS

2.1. Fragmented management of coastal waters 23
2.1.1. The coastline: in-between integrated water management policies . 24
2.1.2. Coastal waters: controlled and regulated according to their uses . . 27
2.2. The difficulty of publicizing coastal water quality 30
2.2.1. The "controlled publicity" of coastal water quality and the consequences on shellfish production in the Pertuis Charentais 31
2.2.2. The quality assessment of variable-geometry coastal waters 34
2.3. The Marine Protected Area: going towards a more coherent management of the "Sea–Land" interface? 37
2.4. Conclusion 38
2.5. References 39

Chapter 3. Comparative Biomarkers Study in Two Scallop Organs to Establish Guidelines for Evaluating French Atlantic Coastline Water Quality 41

Marine Paomia BREITWIESER, Marine BARBARIN, Valérie HUET, Emmanuel DUBILLOT, Marianne GRABER and Hélène THOMAS

3.1. Introduction. 41
3.2. Materials and methods 42
3.2.1. Study sites and sampling collection 42
3.2.2. Biochemical analysis. 44
3.2.3. Data treatments 45
3.3. Results. 45
3.3.1. Superoxide dismutase 47
3.3.2. Glutathione S-transferase 47
3.3.3. Malonyldialdehyde. 47
3.3.4. Laccase. 48
3.4. Discussion 48
3.5. Conclusion 50
3.6. Funding and acknowledgments 50
3.7. References 51

Chapter 4. Tracer Study and *Mimachlamys varia* Bivalve Tools, a Multidisciplinary Guideline to Assess Environmental Quality in the La Rochelle Harbor Over Three Years 55

Marine Paomia BREITWIESER, Benoit SIMON-BOUHET, Jean-Rémy HUGUET, Valérie HUET, Carine CHURLAUD, Marine BARBARIN, Angélique FONTANAUD and Hélène THOMAS

4.1. Introduction. 55
4.2. Materials and methods 58
4.2.1. Field collection and tissue sampling. 58
4.2.2. Tracer study of hydrodynamic processes in the harbor area 59
4.2.3. Trace elements and biomarkers of effect analyses 60
4.2.4. Biochemical biomarkers of exposure analysis 61
4.2.5. Data analysis. 62
4.3. Results. 63
4.3.1. Trace elements and biomarkers of effect analyses 63
4.3.2. Biochemical biomarkers of exposure analysis 67
4.3.3. Relationship between biomarkers of exposure (e.g. SOD, GST, laccase, MDA) and effect (e.g. heavy metals) 71
4.4. Discussion 73
4.4.1. Bioaccumulation 73
4.4.2. Health. 77
4.4.3. Tracer study in the bay of La Rochelle 79
4.5. Conclusion 83
4.6. Funding and acknowledgments 84
4.7. References 84

Chapter 5. ISDAMP Project Dissemination and Continuation for Local Authorities' Oil Spill Preparedness 89

Frédéric MUTTIN and Rose CAMPBELL

5.1. Introduction. 89
5.2. Project results 90
5.3. Project audiences 93
5.3.1. International institutions. 93
5.3.2. National Contact Points 94
5.3.3. Atlantic Arc regional authorities 95
5.3.4. Local authorities in La Rochelle 96
5.3.5. Discussion 98
5.4. Project continuation 98
5.5. Conclusion 101
5.6. Acknowledgments. 101
5.7. References 101

Chapter 6. Environmental Quality of Coastal Areas in the Mediterranean Sea and Potential Risks to Human Health 103

Marion PILLET, Michel MARENGO, Sylvie GOBERT, Pierre LEJEUNE, Michèle LEDUC, Lovina FULLGRABE, Stéphane LE FLOCH and Hélène THOMAS

6.1. Introduction. 103
6.1.1. Chemical pollution in coastal marine areas. 103
6.1.2. Bioindicators in marine coastal water 105
6.1.3. Biomarkers used to monitor pollution. 106
6.2. Situation in the Mediterranean Sea 108
6.2.1. Characteristics of the Mediterranean system 108
6.2.2. Contamination at the lower trophic levels 110
6.3. Mussel biomonitoring in the Mediterranean Sea 111
6.3.1. Biomonitoring of mussels 111
6.3.2. Trace element contamination of mussels in the Mediterranean Sea 112
6.3.3. Organic contamination of mussels in the Mediterranean Sea 114
6.3.4. Physiological functions affected by pollution 117
6.4. Potential human risk assessment 118
6.4.1. Seafood, potential risks and benefits for human health. 118
6.4.2. Potential risks associated with mussel consumption 119
6.5. Perspectives in North Corsica ports 122
6.5.1. Multibiomarker approach 122
6.5.2. The case of harbor areas. 123
6.5.3. The QUAMPO project. 123
6.7. Acknowledgments. 124
6.8. References 124

Chapter 7. Possible Interactions Between Environmental Microplastics, Copepods (*Tigriopus brevicornis*) and Oysters (*Crassostrea gigas*) in the Pertuis Charentais Sea 145

Arno BRINGER, Emmanuel DUBILLOT, Grégoire PRUNIER and Hélène THOMAS

7.1. Introduction. 146
7.2. Materials and methods 147
7.2.1. Study area 147
7.2.2. Experimental design 148
7.2.3. Sampling and observation methods 150
7.3. Results and discussions on the presence of copepods in seawater and in oysters. 150
7.4. General conclusion and perspectives 153
7.5. References 153

Chapter 8. Health Assessment of Juvenile Scallops in the La Rochelle Harbor 157

Marine Paomia BREITWIESER, Marine BARBARIN, Angélique FONTANAUD and Hélène THOMAS

8.1. Introduction. 157
8.2. Assessment methods of the health status of marine bivalves 158
8.2.1. Sampling sites and caging transplantation procedure 158
8.2.2. Water physicochemical assessments. 159
8.2.3. Trace element assessments 159
8.2.4. Biomarker assessments 159
8.2.5. Statistical analysis 160
8.3. Results and discussion on the health status of juvenile scallops 161
8.3.1. Water physical variables. 161
8.3.2. Health biomarker assessments 162
8.3.3. Heavy metals 167
8.3.4. Principal components analysis 171
8.4. References 173

List of Authors 177

Index 181

Preface

Maintaining the quality of marine environments is a tremendous challenge in modern societies. The growth of emerging contaminations is a continuous source of novel questions. Coastal zones have a larger concentration of populations that could suffer from significant levels of pollution if regulations and responses are deficient or insufficient. Research and new knowledge can help decision-makers take valuable action during accidental or chronic water pollution.

Keeping in mind that coastal seas' properties and conditions can range to numerous possibilities. First and foremost, in this book we propose to study the La Rochelle territory along the northeast Atlantic Ocean in the western part of France, but we will also address Portuguese and English waters up to Corsica and the Mediterranean Sea region. We underline the multidisciplinarity of theses in ecotoxicology, biology, mechanics and social sciences. Numerous worldwide organizations, institutions and companies involve both theoretical and operational tools to produce decision aid systems. We scope coastal zones and harbors, and aim to also investigate neighboring high sea regions and economic sectors, such as fisheries, marine cultivations and offshore and water industries.

Throughout eight chapters, we will touch on numerous contaminants like emerging substances and microplastics, the societal dimension together with the environmental issue and, more specifically, the health risks assessment by biomarkers. Some mutual concepts are, for example, water quality management in harbors or the downscaling approaches in numerical models and social sciences.

Chapter 1 presents a European project dedicated to floating oil spill containment booms on the sea surface. We couple numerical models for both coastal hydrodynamics and boom structural analysis. Four experiments in harbors validate our approach. Dedicated to local authorities and operators, our goal is to share numerical modeling results in guides and training.

Chapter 2 describes the correlation and antagonism between the numerous stakeholders and actors. We study the stratification and overlapping of rules, legislations and directives. The social sciences and the coastal water governance are part of the complexity of the water quality challenge. The authority on data and information becomes mandatory for local and national authorities, in order to handle the correct responses to an environmental threat.

From an ecotoxicological point of view, Chapter 3 shows numerous sites (13) in the La Rochelle and Britany areas. We study biological tissues that are able to produce a valuable biomarker. A sentinel species allows one to control and monitor the water quality – the scallop *M. varia* is proposed. The associated multivariate statistical model is a challenging outlook for a smart shoreline environmental assessment.

In Chapter 4, we study the water quality at the level of a harbor, the La Rochelle leisure marina. We establish the link between a biomarker and the contamination of the harbor by hydrocarbons and heavy metals. We process together water flux, and chemical and biological variables upon a sentinel species. This investigation into ecotoxicology allows us to understand the human pressure found in intra-harbor contamination.

Chapter 5 describes the ISDAMP project interests for global to local institutions. Dedicated to local stakeholders, we address the construction of a follow-up project in response to oil spills. We underline the fact that positive attitudes will lead to local behavior changes in both prevention methods and the level of preparedness.

Chapter 6 shows research performed in collaboration with the wonderful STARESO marine station in Corsica, which reviews organic and inorganic pollutions in the Mediterranean region. Bioaccumulation in mollusks can reach critical levels in the food web, affecting human health. The ongoing QUAMPO project will lead to the biomonitoring of the water quality in Mediterranean harbors.

Microplastics pollution is studied in Chapter 7. Addressing a first step in the trophic chain toxification through a benthic copepod, we found the presence of microplastics, which impair health. Performed in the Charente-Maritime region, this research presents results on oyster contamination.

Juvenile scallops' health assessment is addressed in Chapter 8. The difference between adults and juveniles' bioaccumulation processes can be significant for inorganic chemical pollutants. This research improves the knowledge of a sentinel species, dedicated to heavy metals concentration appraisal in harbors.

We use accessible concepts, examples and illustrations to demonstrate the good ecological status of marine waters.

We gratefully thank the authors for their valuable contributions and their carefulness in using complementary approaches.

Frédéric MUTTIN
Hélène THOMAS
June 2021

1

Oil Spill Boom Strategy for Local Authorities Facing Coastal and Harbor Pollutions: A Survey of the European Research Project ISDAMP+

The Charente-Maritime department engineering school, EIGSI, the TU Lisbon engineering school, IST, and Falmouth harbour were partners of the ISDAMP+ project. We present a modeling tool forecasting the hydrodynamic conditions and the structural behavior of floating barriers for the decision aid of local authorities and harbor masters in marine pollution containment. To improve and validate our models, we carried out four experiments in the harbors of Lisbon, Falmouth, Rochefort and La Rochelle. Using model results remains difficult for local responders, however simplification could cause operational issues for numerical models. Sharing experiences and best practices is essential for further progress.

1.1. Introduction

This chapter is dedicated to maritime pollution preparedness, particularly the practice of floating barriers being a core tool to contain oils on the sea surface.

Chapter written by Frédéric MUTTIN and Rose CAMPBELL.

For a color version of all figures in this book, see www.iste.co.uk/muttin/marine.zip

This chapter is split in to five parts. This first one presents the context and project partnership, the second the hydrodynamic model, the third the barrier model, the fourth the coupling between the fluid and solid mechanics and the fifth gives details about the four experiments performed *in situ*.

1.1.1. *Challenge*

We must consider local responses to oil spills through a specific decision aid system. A solely global scale approach appears to be insufficient. A local entity often knows the characteristics of its space very well because it regularly carries out actions there.

The problem addressed is to gather the available data and models efficiently, for example, data regarding the geolocation of pollution, marine currents and the floating barrier behavior. To manage the data, we developed two software tools, named MOHID and BARRIER. To assess the results of these tools, we performed four experiments during the project.

The participating states' policies which are addressed are the marine pollution preparedness area in relation to two topics:

– increasing the preparedness for the coastal consequences of marine pollution accidents [ALV 15];

– using and developing e-learning tools [MAL 19] for spreading knowledge about civil protection and marine pollution.

Our tools could potentially be broadened for other scopes of civil protection, such as coastal flooding.

1.1.2. *Problem*

The following figure illustrates a common question about the best placement of a floating barrier. The answer depends on available mooring possibilities, water current intensities and pollution location and drifting patterns. This barrier device, often called a "boom" and sometimes a "fence" or "curtain", is a mechanical response to contain or deviate floating oils during their entrainment by sea current flow [AMI 09]. The benefit of this device is to concentrate the pollutant and allow for on-sea recovery. Consequently, shoreline pollution is reduced.

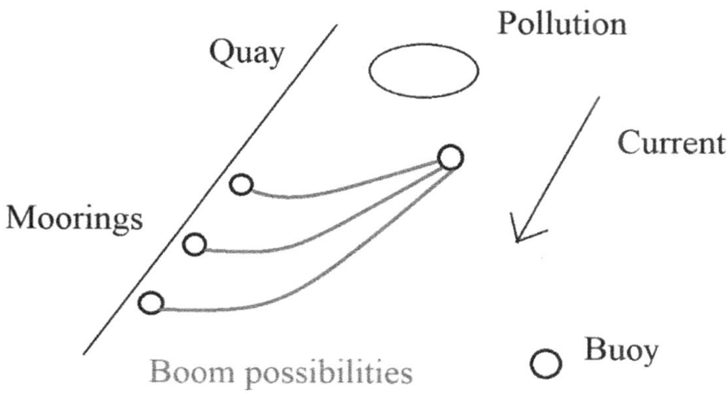

Figure 1.1. *Boom possibilities along a quay in terms of mooring placements, pollution location and current intensities, which themselves are modeled or measured by a buoy*

1.1.3. *Project partners*

This chapter examines an action named ISDAMP+, implemented from 2012 to 2015 in three European countries – France, Portugal, and the United Kingdom – with the European Commission Civil Protection and Humanitarian Aid Office DG ECHO. The project partners are:

– EIGSI: one of the oldest French engineering schools. It establisted Ecole Violet in 1901 in Paris, then in La Rochelle in 1991, and a foreign campus opened in Casablanca, Morocco, in 2005.

– IST: the engineering school of the Technical University of Lisbon, Portugal. It is the largest and the oldest Portuguese engineering school. The MARETEC team specializes in the numerical modeling of aquatic ecosystems, with special emphasis on the hydrodynamics of coastal zones.

– Action Modulers: a spin-off of IST. Its research and development department has worked continuously on the development of the MOHID Water Modeling System and the graphical user interface of MOHID, MOHID Studio.

– Falmouth Harbour Commissioners (FHC), including the Carrick Roads, Cornwall, United Kingdom: a statutory harbor authority, reputed to be the third largest natural harbor in the world. It has practical experience and expertise in oil pollution response. It plans and prepares responses to potential future maritime pollution incidents.

1.2. Coastal hydrodynamic model

This model forecasts ocean parameters and sea-state to estimate the location of oil booms if an oil spill occurs. A hydrodynamic module is the core of the MOHID Water modeling system [BRA 04]. We consider the three-dimensional Navier-Stokes equations under the Boussinesq and hydrostatic approximations [FRI 06]. We adapt the three dimensions of the model with a superposition of two-dimensional surfaces. On each of these parallel sigma layer surfaces (sigma vertical levels) along the water depth, we solve the Saint-Venant equations [AUD 05].

MOHID Studio is a graphical user interface for the MOHID Water modeling system. MOHID Studio enables users to edit data files, create and launch MOHID simulations and analyze model results. MOHID and MOHID Studio can run on desktop computers.

METEO-FRANCE and METEO-GALICIA provide publicly available meteorological data, shown on fine resolution meshes or grids. An operational hydrodynamic model was necessary. We apply the Automatic Running Tool (ART) used at the MARETEC lab. This software tool permits automatic simulations of specific data on the MOHID solver. Not long after, we deemed it necessary to run the atmospheric model Weather Research and Forecasting (WRF), using ART.

Since 2004, MARETEC lab (IST) has implemented an operational wave forecasting system for the North Atlantic and Portuguese coasts, based on the WAVEWATCH III model (version 2.22). The ISDAMP+ project allows setting up such a system for Portuguese and South West French coasts, in order to contain oil leak pollution. We deliver wave data to ensure appropriate conditions and the best locations for floating barrier deployment during emergencies.

We implemented the MOHID hydrodynamic model for the La Rochelle region, France [ASC 15], in 2014. The coastal hydrodynamic depends on the flows of the rivers and estuaries of the Sèvre-Niortaise, the Charente, the Seudre and the Gironde [RAY 15]. The unstructured grid resolutions range from approximatively 50 km in the Bay of Biscay grid, to 400–200 m in the

La Rochelle Bay fine local grid. A downscaling approach complies with a multigrid scheme. The imbedded mesh grids have successively finer resolutions. It allows for a proper representation of ocean dynamics and wave propagation. The calibration and validation of model results use open data registered by buoys, stations and cruise campaigns. Bathymetry, global ocean circulation and principal tidal constituents give the boundary conditions and an initial state of the fluid velocity and water depth.

1.3. Boom structural analysis model

EIGSI has produced a software named BARRIER, which provides a decision support system for oil spill boom placement. These depend especially on water current. A map displays the current data and results. Software options permit two kinds of booms, air inflated float curtains and solid foam flotation barriers or fences. The numerical method implemented in BARRIER follows a multigrid and multimodel scheme. It proposes four kinds of mesh. Each of them considers a mechanical model with a low or high refinement of the boom geometry. The model names are BAR0D, BAR1D, BAR2D and BAR3D. They assimilate a strained boom geometry as a straight line, a catenary curve, a 2D curvilinear finite-element domain, named cable, and a 3D tensile finite-element surface, named membrane, respectively.

The displacement u and tension T of a 2D cable domain define the function (u,T)(U), written in terms of free-surface velocity U. This one with the water depth h solves the Saint-Venant equations associated with the sea surface layer.

The first operational test cases reveal that BAR0D and BAR1D give a restricted prediction of the boom geometry and containment efficiency [GON 14]. However, the lack of indication on boom tension T, upsets the decision aid for mooring definition. Another outcome of BAR0D or BAR1D is a boom initial position for the non-linear iterative solvers included in BAR2D and BAR3D. The results given by BAR2D are valuable when it comes to proposing a boom deployed geometry in a given environment. BAR2D needs only to moderate the computational cost. Updating the contingency plan is fast and simple.

From the ISDAMP+ point of view, the Rochefort and La Rochelle Chef de Baie experiment computations only use BAR1D and BAR2D results. The BAR3D model is more consistent with a research activity. Its interface with the convergence of the iterative solver remains difficult to handle for end-users. The Departmental Directorate of Territories and the Sea (DDTM, *Direction Departementale des Territoires et de la Mer*) of Charente-Maritime recommends that these models need to request minimal filling of inputs, variables, parameters, results and outputs. The model should thus become operational.

A useful BAR3D output is a stress map of the fabric material along each boom section. The 3D membrane geometry exhibits the underwater boom skirt. The vertical skirt angle is a main indicator of the oil containment efficiency of the boom.

After an incident on the Maurienne River (French Alps) in November 2014, we have applied BARRIER to analyze the failure of a boom moored in an artificial lake. By using the open data of the hydropower reservoir and channel, the outputs of BARRIER could improve future strategies.

1.4. Hydrodynamic and structural models coupling

At the end of the project, after a battery of tests, the MOHID and BAR2D models were coupled with general oceanographic conditions. An Open-MI architecture links these models. The MOHID model can use fine spatial resolutions within a range, from 30 to 100 m. The BAR2D model generally considers boom sections under 100 m in length. The boom axial resolution ranges from 1 to 10 m. The ocean and boom meshes have different spatial resolutions. An interpolation of the velocity field of the sea current allows the models exchange. The hydrodynamic pressure on boom finite-elements stems from the velocity set on the sea-surface finite-volumes.

Based on responders' feedback and the functional possibilities of the numerical coupling, we integrate several external factors bearing on boom deployment. The general architecture of the system incorporates weather, river discharge rates, boom length, mooring locations and initial oil slick area.

For example, we consider the inner Lisbon estuary. We position an oil spill and an 800 m long barrier moored with five anchors. It used the IST Tagus operational model implemented previously. Figure 1.2 shows the shoreline, sea current, boom geometry and oil spill, four hours after it began. We see that the current faces the four boom sections and that oil particles are trapped [FEN 19a]. They do not enter the estuary branch.

Figure 1.2. *The coupling of MOHID and BAR2D implemented in the Tagus river, with an oil spill and a boom in the inner Lisbon estuary*

The map of the figure comes from a WMS Server (Web Map Service interface standard). WMS is a standard of the Open Geospatial Consortium (OGC).

Over time, we update the boom mesh after a given number of hydrodynamic time steps. A Lagrangian scheme is used to update the oil particle positions. When an oil particle tackles the boom, under a relative normal velocity of 0.35 $m.s^{-1}$, its updated velocity is zero. Over this threshold, the oil particle flows below the boom. The oil spill model (OSM) implemented in MOHID is not detailed here [FER 14].

1.5. Full-scale experiments

We present the four experiments performed over the course of the project. These were carried out to validate the models and for training purposes, using different floating barriers in Falmouth, Lisbon, Rochefort and La Rochelle Chef de Baie; all being North East Atlantic harbors.

1.5.1. *Falmouth*

The Falmouth experiment, named Exercise Maia, took place in May 2014 in front of the Falmouth docks within the Carrick Roads, between 50° 09.7247 N, 05° 03.1150 W and at the southern end 50° 09.6047 N, 05° 03.0343 W. The simulation was an oil spill starting at 05:30 am on May 12th, 2014. The pollutant amount was approximately 4 tons of intermediate fuel oil. The scenario was a collision between a stationary bunker barge [RED 02] and a vessel calling for bunkers in the Carrick Roads area. The collision resulted in structural damage to the bunker barge and oil released as pollution in that location.

The boom used several instruments, GPS tracers, load shackles and buoy data, and was positioned using PISCES numerical simulations. The boom geometry is a 280 m long chevron. FHC deployed 17 lengths of 75 cm inflated depth by 20 m length PU alloy inflatable boom manufactured as a Sea Sentinel Boom (see Figure 1.3).

Figure 1.3. *Exercise Maia in the Falmouth estuary May 2014, Falmouth Harbour Commissioners*

FHC placed battery operated GPS tracers regularly spaced along the boom. The autonomous satellite-based tracers were located on the ends of the two arms of the boom and in the center of both, with six tracers in all. These would report the boom position every 30 minutes and allow a "real time" web-based visualization of its movement. The existing data buoy measured meteorological and tidal information. It was positioned in the center of the mouth of the boom, and was able to demonstrate a direct correlation between boom movement, wind speed and direction and tidal flow. It was used for the real tidal predicated data for that day, but FHC made up the wind information.

FHC also moored two load shackles. The horizontal chain sections clamp on the two main moorings containing the boom load cells. They fitted the seabed. The same screen showed the data and the boom position made using the tracers, giving a complete dynamic picture of position, wind, tide and chain loads. Also available was a visualization from FHC's oil simulation model, made using PISCES, showing the V-shape boom location in the harbor.

The exercise was not exactly a real time exercise because the various stages of the response were staggered over a period of a few days. The exercise figured out the UK Maritime Coastguard Agency (MCA) Oil Spill Contingency Planning (OSCP) guidelines' obligations.

We discovered the moorings logs in Figure 1.4, given the sea current speed, the tide height and the mooring loads. Tide height ranges from 0.37 m to 5.25 m over the course of two days. Current speed reaches 0.68 knot, 0.35 $m.s^{-1}$, its threshold for an efficient oil containment by floating boom [SHI 20]. Mooring load attains 2.09 tons, taken as a limit for granite blocks' adherence on the seabed.

FHC realized a risk and vulnerability assessment and a modeling. The boom installation duration was nine days. Exercise Maia was ambitious with a strong training issue. We edited reports and comparison of the exercise and that of other partners. We realized a training film on the exercise for local operators and another training film for high-level authority.

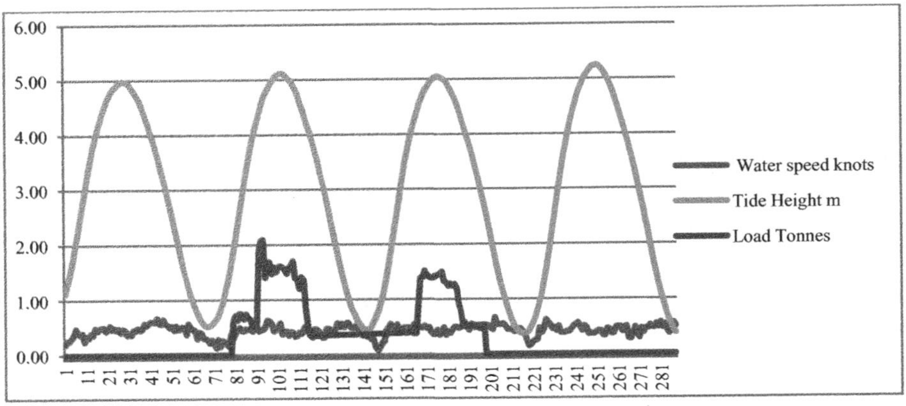

Figure 1.4. *Logs of current speed, tide height and mooring load of the V boom, Falmouth Harbour Commissioners*

Falmouth Harbour Commissioners organized an all-day Falmouth Oil Spill Response Conference in Falmouth, Cornwall (UK) on May 21st, 2014. The European projects SPRES (Atlantic area transnational program, ERDF) and ISDAMP+ helped to share the best practices for oil spill responders and scientific spill response tools for academics and researchers. The conference had two sessions:

– a morning for sharing the best practices from oil spill responders (National and European spill responses, real-life booming exercises, experiences and lessons learnt from response contractors);

– an afternoon for displaying innovative technical and scientific spill response tools (spill fate modeling, environmental vulnerability assessments, and software tools).

1.5.2. *Lisbon*

The Lisbon ISDAMP+ experiment occurred October 14th and 15th, 2014 in the Tagus estuary, 38.678624° N and 9.195935° W. The exercise, coordinated by IST, included the Lisbon Port Administration (APL) and the REPSOL company. When choosing the exercise location, the presence of liquid bulk terminals, their strategic position, model accuracy, harbor traffic, and protection objectives were considered [OLI 14]. This test case used booms as physical barriers, giving a closed loop around a wreck. The boom referenced NOFI 350 EP and was 150 m long. The boom mooring used two bollards on the Banática Liquid Terminal. It allowed us to study the behavior of those barriers during an accident. Three persons from the APL made the operation, mobilizing an APL ship. During the exercise, we registered the boom deformation manually, using photos (see Figure 1.5) and automatically, using three drifting buoys attached to the boom. On October 14th, 2014, the water height was 1.16 m at 1:13 am, 3.25 m at 7:50 am, 1.2 m at 1:48 pm and 2.98 m at 8:20 pm.

Figure 1.5. *Exercise at REPSOL terminal in October 2014, Lisbon, Instituto Superior Técnico*

We released a set of four drifting buoys to test and validate the ISDAMP+ hydrodynamic model using the MOHID 100 m resolution grid implemented in the study area. These buoys followed the surface hydrodynamic field as Lagrangian tracers. We compared buoy drift measurements with model results for numerical tracers. The aim was also to locally measure the flow speed with a current meter, the surface wind with a manual anemometer and the water's physical parameters, salinity, temperature and turbidity [ROD 17] with a multi-parametric sounder.

1.5.3. *Rochefort*

An ISDAMP+ boom deployment experiment took place in basin number 3, at dock number 6, of the commercial port of Rochefort Tonnay-Charente, 45.950721 N, -0.954089 E. Rochefort is one of the most important ports in the Charente-Maritime department in the western part of France. Located within the marshy Charente estuary, at 18 km of the Pertuis d'Antioche, and the Bay of Biscay, Rochefort is a medium-sized river port.

The scenario focusses on providing a response solution in the case of a small portuary accidental release. For example, an incident occurred during bunkering from a truck, or during the breakdown of an old ship [GIR 04], and a small amount of oil polluted the port waters. The pollutant is less than 1 ton of intermediate fuel oil from slop or bilge waters. EIGSI engineering school has performed the boom experiment from both operational and scientific points of view. EIGSI purchased a 50 m long curtain boom to that end. The boom reference is EUROSORB 1830 BCR, with a total height 0.48 m, draught 0.3 m, freeboard 0.18 m and chain mass 3.2 $kg.m^{-1}$.

Each basin sweeping run consisted of a small tug towing a boom end-point from the quay to its opposite in the basin and back. The other end-point of the boom remains fixed to the quay. The tugboat was fitted with a GPS tracer, which provided the bearing and position of the boat during each run. The different runs were on December 5th, 2014, from 09:30 to 13:00 and 14:00 to 16:00 (see Figure 1.6).

The influence of tides on the basin hydrodynamic is negligible. Gates closed the basin of the harbor. Some key facts of the experiment were:

– C-shape boom deployment, held in place using a port ladder and a boat;

– boom deployed in the water from a quay, providing a dynamic containment and concentration of the pollutant;

– 13 people involved.

Figure 1.6. *Sweeping boom in the port of Rochefort, December 5th, 2014, EIGSI*

The main goals of the experiment were to:

– observe the boom trajectory when towed at different angles and speeds;

– measure the tension on a towed boom with a dynamometer on the ladder;

– perform a BAR2D dynamic simulation, accurately reproducing what was observed in real-life;

– compare simulated and observed tension values.

If an oil, chemical or HNS pollution occurs, the SDIS-17, the city fire brigade, intervenes at the request of the harbor master. The operations rarely require a model.

1.5.4. *La Rochelle Chef de Baie*

The Chef de Baie harbor is a fishing port located in the city of La Rochelle, under the policy of the Charente-Maritime council. The area of the pollution scenario 46.149724 N, -1.223472 E is a place of refuge dedicated to small fishing vessels less than 12 m long.

The experiment occurred on March 16th, 2015. Two sections of containment boom, 10 m long each, and one section of sorbent boom surround the pollution slick. The sorbent section was 15 m long and constituted the gate of the entrance for a leaky ship. The experiment duration was 6 h, +-3 h around full tide.

The mooring of this triangular boom plan is between the fueling station, schematized on the left of Figure 1.7, and the mobile crane quay, on the right of the figure. The arrow at the bottom of the figure represents a forcing current coming from the south, and the harbor entrance. The forcing can be disrupted by tidal eddies, wave refraction and wind [FEN 19b].

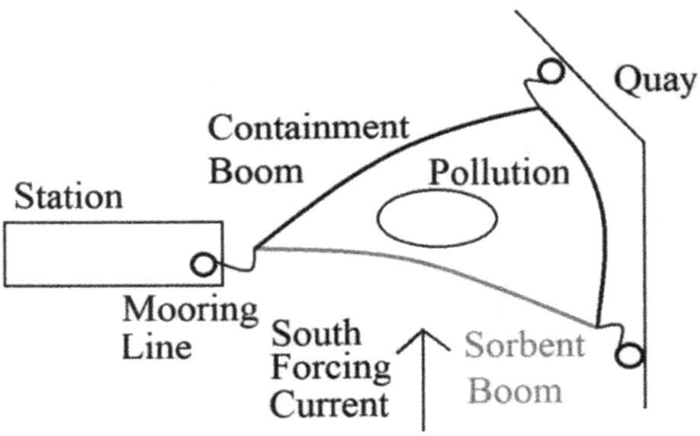

Figure 1.7. *Simulating containment of a leaky ship in the place of refuge of La Rochelle fishing harbor, EIGSI*

Figure 1.8 shows a view between the floating pontoon (on the right) and the handling quay (on the left).

The direct opening of the port into the ocean necessitates a large quay wall height that renders difficult antipollution operations above them.

The key facts of the experiment are the following:

– the forcing direction could change significantly if a squall occurs;

– in the west and north directions, atmospheric and oceanic forcings are not directly taken into account;

– the wave agitation in the harbor could be significant, with a north-west oceanic wave source;

– boom deployment was made by two operators;

– no boats were used.

Figure 1.8. *Experiment at Chef de Baie fishing harbor, March 2015, EIGSI*

In the context of the experiment's preparation, the 6th Oil Spill Studies Workshop was held at EIGSI La Rochelle on March 11th, 2015. Three sessions were held on:

– ISDAMP+ software results based on models;

– coastal sensitivity, ecotoxicology analysis and juridical concerns;

– experiments and case studies in ports and harbors.

The Prefect of Charente-Maritime has the authority to manage pollution in the harbor. For oil pollution from outside, hence the ocean, a contingency plan, POLMAR, describes the harbor protection by booms. This plan uses the FORBAR 2D structural model for booms.

1.6. Conclusion

The following table gives an overview of the models and software used during the ISDAMP+ experiments. The hydrodynamic models provide high quality 3D results, including calibrated oil drifting and optionally oil weathering processes. The model dimensions for boom structural analysis can be in 1D, 2D or 3D. The coupling between flow and structure is operational [ŁAZ 13].

Models and software	Falmouth Harbour	Lisbon terminal	Rochefort Harbor	Chef de Baie Harbor
Coastal hydrodynamic	DELFT3D	MOHID3D	NO	MOHID
				SWAN
Boom structural analysis	PISCES II	NO	BAR2D	BAR2D
			Dynamic	
Oceanic flow and boom	PISCES	MOHID	NO	NO
Structure coupling		BAR2D		

Table 1.1. *Summary of models and software used during the four ISDAMP+ experiments*

We have made significant progress in the prediction of oceanic parameters in the Bay of Biscay and the English Channel. Valuable tools propose coarse and fine meshes for the structural analysis of oil spill booms. Level 1 (Tier 1) exercises were tried out by local operators and revealed positive aspects of the research carried out to authorities and decision-makers. Nevertheless, scientific and pollution response works remain parallel and distinct activities. A continuous effort of exchange gives synergies between them. The number of pollution configurations is so large that only a dialog on best practices allows one to share their management. We experimented with the collaboration of national and local authorities.

The project's aims were to gather virtual operational tools with real-life experiments. Both activities have their intrinsic accomplishments. Generally, the tools and results implemented by the project are well-evaluated by the partners themselves. The project results benefitted from external assessments that advise a broad dissemination of them.

The effective usage of the software products needs close cooperation between researchers and software developers. The distinction between the end users and the donor or funder is prominent for the tools' adoption by the national authorities in charge of accidental pollution response.

The actual expansion of data sharing and interoperability in the maritime sector could be a main driver for the future. Collaboration in a virtual environment appears as a main evolution of decision aid tools for preparedness in maritime pollution.

1.7. Acknowledgments

We are grateful for the support of the European Union "Humanitarian Aid and Civil Protection", DG-ECHO, and the funding of the ISDAMP+ project by the Civil Protection Financial Instrument under grant no. 638516/2012/ECHO/A5/SUB.

1.8. References

[ALV 15] ALVES T.M., KOKINOU E., ZODIATIS G. *et al.*, "Modeling of oil spills in confined maritime basins: The case for early response in the Eastern Mediterranean Sea", *Environmental Pollution*, vol. 206, pp. 390–399, 2015.

[AMI 09] AMINI A., SCHLEISS A.J., "Numerical modeling of oil-water multiphase flow contained by an oil spill barrier", *Engineering Applications of Computational Fluid Mechanics*, vol. 3, no. 2, pp. 207–219, 2009.

[ASC 15] ASCIONE KENOV I., MUTTIN F., CAMPBELL R. *et al.*, "Water fluxes and renewal rates at Pertuis d'Antioche/Marennes-Oléron Bay, France", *Estuarine Coastal and Shelf Science*, vol. 167(A), pp. 32–44, 2015.

[AUD 05] AUDUSSE E., "A multilayer Saint-Venant model: Derivation and numerical validation", *Discrete and Continuous Dynamical Systems – Series B*, vol. 5, no. 2, pp. 189–214, 2005.

[BRA 04] BRAUNSCHWEIG F., LEITAO P.C., FERNANDES L. *et al.*, "The object oriented design of the integrated Water Modeling System MOHID", *Developments in Water Science*, vol. 55, no. 2, pp. 1079–1090, 2004.

[FEN 19a] FENG X., SHI Y., "Numerical study of containment of spilled medium-viscosity oil in wave–current flow", *Journal of Engineering Mechanics*, vol. 145, no. 8, p. 04019056, 2019.

[FEN 19b] FENG D., HODGES B.R., SOCOLOFSKY S.A. *et al.*, "Tidal eddies at a narrow channel inlet in operational oil spill models", *Marine Pollution Bulletin*, vol. 140, pp. 374–387, 2019.

[FER 14] FERNANDES R., LOURENÇO F., BRAUNSCHWEIG F. *et al.*, "Dynamic risk assessment of shoreline contamination from ships: Integrating an oil spill model", *International Oil Spill Conference Proceedings*, vol. 2014, no. 1, p. 299678, Savannah, USA, May 5–8, 2014.

[FRI 06] FRINGER O.B., GERRITSEN M., STREET R.L., "An unstructured-grid, finite-volume, nonhydrostatic, parallel coastal ocean simulator", *Ocean Modelling*, vol. 14, nos 3–4, pp. 139–173, 2006.

[GIR 04] GIRIN M., "European experience in response to potentially polluting shipwrecks", *Marine Technology Society Journal*, vol. 38, no. 3, pp. 21–25, 2004.

[GON 14] GONG K., TKALICH P., XU H., "The numerical investigation on oil slick behavior behind the oil boom", *Journal of Environmental Protection*, vol. 5, no. 8, pp. 739–744, 2014.

[ŁAZ 13] ŁAZUGA K., GUCMA L., PERKOVIC M., "M/t "Baltic Carrier" accident. The reconstruction of oil spill with PISCES II simulator application", *Zeszyty Naukowe/Akademia Morska w Szczecinie*, vol. 36, no. 108, z.1, pp. 110–115, 2013.

[MAL 19] MALLAM S.C., NAZIR S., RENGANAYAGALU S.K., "Rethinking maritime education, training, and operations in the digital era: Applications for emerging immersive technologies", *Journal of Marine Science and Engineering*, vol. 7, no. 12, p. 428, 2019.

[OLI 14] OLIVEIRA E.R., SILVEIRA B., ALVES F.L., "Support mechanisms for oil spill accident response in costal lagoon areas (Ria de Aveiro, Portugal)", *Journal of Sea Research*, vol. 93, pp. 112–117, 2014.

[RAY 15] RAYSON M.D., GROSS E.S., FRINGER O.B., "Modeling the tidal and sub-tidal hydrodynamics in a shallow, micro-tidal estuary", *Ocean Modelling*, vol. 89, pp. 29–44, 2015.

[RED 02] REDDY C.M., EGLINTON T.I., HOUNSHELL A. *et al.*, "The west Falmouth oil spill after thirty years: The persistence of petroleum hydrocarbons in marsh sediments", *Environmental Science and Technology*, vol. 36, no. 22, pp. 4754–4760, 2002.

[ROD 17] RODRIGUES M., FORTUNATO A.B., "Assessment of a three-dimensional baroclinic circulation model of the Tagus estuary (Portugal)", *AIMS Environmental Science*, available at: https://doi.org/10.3934/environsci.2017.6.763, 2017.

[SHI 20] SHI Y., WEI J., LI S. *et al.*, "Experimental study on containment of moderate-viscous oil by floating boom subject to waves and currents", *Applied Ocean Research*, vol. 94, p. 102003, 2020.

2

Governance and Public Policies of Coastal Water Quality in the Pertuis Charentais

The coastal area of the Charente-Maritime region is highly attractive and is mainly used for economic and recreational activities. Each year, three million tourists visit this territory, which stretches across 460 kilometers of coastline. According to the National Statistics Institute, more than 6.5% of the local economy relies on tourism. The Pertuis Charentais, which are the receptacle of three rivers (the Sèvre-Niortaise, the Charente and the Seudre) create a good natural environment for shellfish farming. The shellfish sector is an additional structuring activity. It shapes land and sea territories and contributes to the development of department identity. In terms of economics, shellfish farming achieves a turnover of 350 million euros and generates nearly 25,000 direct jobs. Around 23% (50,000 metric tons) out of the 217,000 metric tons of shellfish sold in 2016 were shipped by companies located in this department.

Given the importance of its coastal activities, the water quality is of major importance and represents a challenge for the Charente-Maritime. Poor water quality is detrimental to biodiversity. Poor conditions will thus affect the water, its usage and the various environments. People's health and safety may also be impacted. If a potentially toxic bacteria or microscopic algae

---

Chapter written by Alice MAZEAUD, François BÉNITEZ, Anaïs RIDEL, Arno BRINGER, Christine PLUMEJEAUD, Pascal BRUNELLO, Adélaïde LEREBOURS, Yann RABUTEAU and Hélène THOMAS.

For a color version of all figures in this book, see www.iste.co.uk/muttin/marine.zip

was likely to develop, bathing could become dangerous and be prohibited. Poor chemical status could lead to the closure of shore fishing areas, impacting both professional and recreational fishing. Bacteriological contamination could affect aquaculture, which may mean the coastal shellfish cannot be sold. This kind of degradation poses a serious threat to the economic activity of the Charente-Maritime coastline. Thus, it is essential to identify the public stakeholders managing the coastal water quality[1].

In 2017, the Nouvelle-Aquitaine Regional Economic, Social and Environmental Council (*Conseil Economique, Social et Environnemental Régional* (CESER))[2] published, for the first time ever, a report entitled "The Coastal Water Quality in Nouvelle-Aquitaine: Current Status and Prospects", which highlighted the governance complexities. Public policies dealing with this matter are highly fragmented. Despite abundant discourse and strategies promoting Integrated Coastal Zone Management (ICZM) [DRO 13], coastal waters, which are "shallow water bodies close to the coastline"[3], are not a category included in public action policies. Using a geographical definition, coastal waters are not subject to integrated management. Indeed, they are in-between integrated freshwater and integrated maritime water management policies. Usually defined by their uses, coastal waters are divided into bathing waters, shellfish waters, etc. Thus, while European directives focus on achieving "good status", water quality is monitored and controlled based on its uses – bathing, shellfish farming, fishing, etc. – and on public health matters.

There is a need to point out why the degradation of nearshore water quality and its consequences, mainly to shellfish aquaculture, should not involve the public sphere [GAR 10]. Given the importance of socio-economic issues, it may seem surprising that it is not at the heart of both the political and media agenda. Several factors could explain the "controlled publicity" [CAN 12] of coastal water quality. In addition to

1 The analysis is based on documentary analyses and interviews carried out within the framework of two Master's studies interns, F. Bénitez and A. Ridel, funded by the EVOLPEC, *Evolution du littoral sous contrainte naturelle et anthropique des Pertuis Charentais* project.

2 CESER is a regional consultative assembly composed of economic and social actors representing civil society.

3 As there is no official definition, we have used the one provided by CESER in 2017.

having a mediatization strategy that could be tricky, there are difficulties in pointing out who is responsible for pollution. This is mainly due to a lack of scientific expertise and the unbalanced powers among actors working upstream. Finally, there is a need to define and measure quality. Indeed, several regulations define quality thresholds based on pre-identified indicators. But these different quality measurements blur the true notion of quality. Despite positive assessments of nearshore water quality, the increased mortality of oysters demonstrates that current indicators are not perfect. In fact, only chemical or biological elements identified as degradation factors are measured. Therefore, the impact of the recent creation of the Gironde Estuary and Pertuis Marine Protected Area on coastal water quality management is questionable.

2.1. Fragmented management of coastal waters

For several decades, water policies have globally been relying on territories. The catchment area, which is a natural unit, was created to support this policy on territorial management [BRU 11]. In France, these principles were enforced by the 1964 Water Act. They were then integrated into the European Water Framework Directive (WFD) in 2000[4]. The objective is to carry out action plans based on the basin or sub-basin of a lake or river, and not based on their politico-administrative unit. Thus, a hydro-geographically coherent policy should be developed and implemented. Indeed, water flows naturally and acts independently of political-administrative borders or public matters that differ from one catchment area to another (such as floods, pollution and water shortages). This principle was adopted by the Marine Strategic Framework Directive (MSFD) in 2008[5] for maritime water management. Nonetheless, despite promoting ICZM, it has never been implemented. Although counting on territories helps to organize water management (also called the "great water cycle"), sectoral breakdowns based on pre-existing national public actions are ongoing. This includes drinking water and distribution management, flood risk management, bathing waters, waste waters, shellfish waters, etc. This territorial model does not replace the intervention of political and

4 Directive 2000/660/EC of the European Parliament and of the Council from October 23, 2000 establishing a framework for a community water policy.

5 Directive 2008/56/EC of the European Parliament and of the Council from June 17, 2008 known as the "Marine Framework Directive".

administrative authorities. These elements contribute to fragmented coastal water management.

2.1.1. *The coastline: in-between integrated water management policies*

The Water Framework Directive or WFD (2000/60/EC) aims to achieve or maintain a good status of surface water and groundwater within European territories; this includes the outermost regions (ORs). In France, the directive has been implemented on river basins by Water Development and Management Master Plans. The Marine Strategic Framework Directive MSFD (2008/56/EC) aims to achieve or maintain a good environmental status in marine waters throughout Europe, excluding the outermost regions. In France, this was implemented across marine sub-regions from 2012, with the development of Marine Environment Action Plans. According to the government, "these two directives have a common objective of achieving good status of waters to which they apply, waters which partly overlap". Indeed, the coastal waters are overlapping areas. Using the WFD terminology, inshore waters include two components: coastal waters which are "surface waters located between the baseline used to measure the width of territorial waters and a distance of one nautical mile (1852 m)", and transitional waters which are "near river mouths and rivers, partially saline because of their proximity to coastal waters, but fundamentally influenced by freshwater currents". The MSFD includes coastal waters and applies from the baseline.

The goal of these directives is to bring the degradation of water quality to a halt and restore the good status of both land and marine waters. According to the WFD, good status water is defined as "water that allows rich and diversified animal and plant life, that is free of toxic products and available in sufficient quantity to meet all human uses and activities". "Ecological status" looks at the ecosystem as a whole. It is based on biological parameters (such as the abundance of fish species in a river) and considers the physico-chemical parameters (for example the oxygen dissolved in the water or the temperature) and the morphology and hydrology of the environment. Chemical status is assessed on the basis of the presence and concentration of pollutants in water. Chemical status is divided into two categories: "good" and "not good". Ecological status has five categories ranging from "very good" to "poor". Marine waters that achieve good environmental status can be defined as those that "provide ecologically diverse and dynamic oceans and seas, which are clean, healthy and productive within their intrinsic conditions, and the use of the marine environment is at a level that is sustainable, thus safeguarding the potential for uses and activities by current and future generations"; a definition which reflects the above. Both policies set objectives imposing obligations to achieve results according to three principles: integrated management, consideration of environmental and socio-economic data and public participation. The goal is to articulate expertise and management at the river basin level. Following an iterative process, the principles are part of a six-year cycle, which includes successive phases of evaluation, definition and implementation of measures. The WFD predicted that good status would be achieved in 2015, while the MSFD planned to reach it in 2020. Both policies are at different stages of the program, the WFD being in the evaluation phase of the second cycle (2016–2021) and the MSFD having just completed its first cycle (2012–2018).

The policies are implemented in hydro-geographical basins, as shown in Figure 2.1. According to the WFD, coastal waters are connected to the nearest river basin district. The Pertuis Charentais depend on the Adour-Garonne basin for the Seudre and Charente basins, and the Loire-Bretagne basin for the Sèvre-Niortaise basin. These three Pertuis are also located in the river catchment area to which they connect. According to the MSFD, the Pertuis Charentais belong to the submarine region of the Bay of Biscay (see Figure 2.1).

Giving territories the responsibility of carrying public action implies new governance principles, including concerted management. Water policies are a symbol of a participatory shift in public action [PAR 18]. It has long promoted a principle of concerted management involving all water bodies (often referred to as the "Water Parliament" in France). For each basin, an administrative agency, known as the Water Agency, is responsible for implementing the Water Development and Management Scheme. The scheme is defined by a basin committee composed of representatives of the state services, local authorities and water bodies. A local water commission is in charge of defining actions in compliance with the Water Development and Management Scheme and detailed in the River Basin Management Plans. The composition of the commission is defined by law[6] and made to represent all water users. Thus, 50% are representatives of the

6 Article R212-30 of the Environmental Code.

municipalities, 25% are users and 25% are members of the state services. All three Pertuis Charentais are included in the scope of one River Basin Management Plan[7].

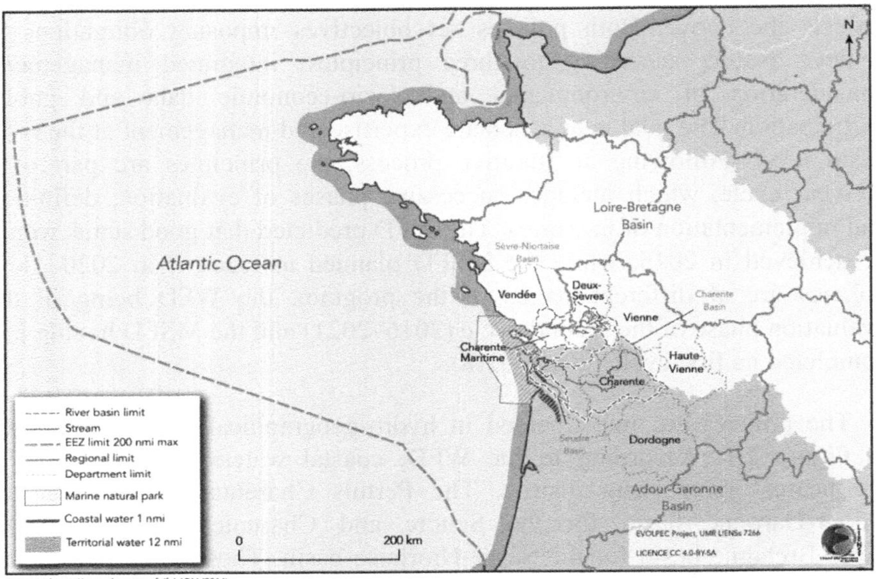

Sources : https://www.data.gouv.fr/fr/, IGN (2014)

Figure 2.1. *Water management territories in the Pertuis Charentais*

Regulations for marine waters are more recent and rely on government services. Indeed, each marine sub-region is led by the Maritime Prefect, responsible for the development and implementation of a reviewable Marine Environment Action Plan, such as the Master Plans for Water Development and Management, every six years. The Prefect is helped by a frontline maritime advisory council, which includes representatives of the state, of public bodies and of local authorities. It also comprises members of professional activities, companies and employee representatives whose actions are directly linked to the exploitation of the sea and the coastline. In

7 Refer to the SAGE Sèvre-Niortaise and Marais Poitevin documentation, which was approved on April 29, 2011 (https://www.sevre-niortaise.fr/documentation/download-category/sage-sevre-niortaise-et-marais-poitevin/), SAGE Seudre approved on February 7, 2018 (http://www.sageseudre.fr/sage), and SAGE Charente pending approval (http:// www.fleuve-charente.net/domaines/le-sage/projet-2).

addition, the advisory council welcomes NGO representatives promoting the protection of coastal and marine environments, or protecting the coastline and sea users.

Coastal waters are interfaces between the two policies (the WFD and the MSFD), which are in-between freshwater and marine water policies. Both management plans should be implemented simultaneously and in a coherent manner. Depending on the impact location and the pressure origin, the Marine Environment Action Plan and/or the Water Development and Management Scheme should apply. The monitoring of water status is an environmental innovation as it aims to preserve good chemical and ecological statuses of water, regardless of its usage. Good status underlines the environmental state and ability to ensure the development of flora and fauna. However, it is not a qualitative marker.

2.1.2. *Coastal waters: controlled and regulated according to their uses*

Historically, water policies are first and foremost public health policies [BAR 19]. States have monitored and regulated water quality to avoid human risk. The promotion of water management organized in territories kept the pre-existing sectoral regulations. Monitoring the ecological status of the water did not substitute, but complemented health monitoring. Thus, prior to the creation of the aforementioned directives (WFD and MSFD), coastal waters and the two activities that are the focus of this chapter – namely bathing and shellfish farming – were regulated, including by European law. The Swimming Directive was adopted in 1976[8] and then replaced by a new one in 2006. It mentions a set of parameters to follow so that bathing waters are compliant. In the Pertuis Charentais, the quality of the bathing water is monitored by a decentralized service of the Ministry of Health, the Regional Health Agency (RHA). In 2018, 91 sea bathing areas were monitored. The majority of the Charente-Maritime sites are of excellent quality. Indeed, 71% of sites are assessed as excellent, 26% as good, 2% as of sufficient quality and one site is of insufficient quality. When quality is insufficient, the mayor should prohibit bathing.

8 Council Directive 76/1160/CCEE of December 8, 1975 on bathing water quality.

The Shellfish Water Directive, adopted in 1979, was replaced by a new directive in 2006[9] requiring States to set threshold values for water quality, establish pollution reduction plans and monitor waters. Article 1 defines shellfish waters "as coastal and brackish waters designated by Member States as in need of protection or improvement to ensure the life and growth of shellfish species (bivalve and gastropod mollusks), thus contributing to the good quality of edible shellfish products". This shellfish directive was repealed in 2013 by the WFD. The biological and physico-chemical parameters of the 2006 Shellfish Water Directive are now used to assess the quality within river basins. Nowadays, the shellfish water safety classification is based on European regulations 853/2004/EC and 854/2004/EC. A decree from 2013[10] enables microbiological thresholds to be set for each category – A, B, C or NC – and for potential recreational and professional fishing activity to be classified. The decree distinguishes three groups of shellfish species (marine gastropods, burrowing bivalves and non-burrowing bivalves), which are classified based on their contamination and purification potential with regard to microbiological contaminants.

Monitoring is carried out regularly by a public state institution, the *Institut Français de recherche pour l'exploitation de la Mer* (IFREMER) (French Research Institute for the Exploitation of the Sea) is in charge of several monitoring networks including the microbiological control network, which monitors the professional production areas to classify or declassify the products (A, B, C, NC); the phytoplankton and phycotoxins observation network, which monitors species that are toxic or harmful to the marine fauna in coastal waters, particularly those eaten by consumers. The chemical contamination observation network assesses the level of chemical contamination in classified shellfish production areas. Based on the chemical contamination observation network assessment, the Departmental Directorate of Territories and the Sea of the Ministry of Ecological Transition and Agriculture classifies areas[11] and triggers alert processes in the event of contamination. In 2017, "the alert system" was used 21 times after contaminant detection and once for preventive reasons (N0), subsequently

9 Council Directive 79/9923/CCEE of October 30, 1979 on the quality required for shellfish waters. Directive 2006/1113/CCE of the European Parliament and of the Council from December 12, 2006 on the quality required for shellfish waters.

10 Decree of November 6, 2013 on the classification for health monitoring and management of production areas and relaying areas for live shellfish.

11 See: http://www.atlas-sanitaire-coquillages.fr/classements-sanitaires.

engaging the level 2 alert system. In 2017, the system led to 41 additional samples and analyses being taken. It is important to note that these devices are only used for professional fishing. In addition, since 1996, in the Charente-Maritime, the regional health agency (RHA) has also carried out bacteriological monitoring for recreational fishing, which resulted in regulatory measures being put into place.

Monitoring the quality of shellfish or bathing waters induces regulatory effects. Bathing, shore fishing or marketing shellfish products may be prohibited due to poor water quality. Given the socio-economic challenges with regards to tourism and shellfish farming activities, local authorities may also decide to pursue proactive policies. They directly or indirectly contribute to supporting tourism and shellfish activities, particularly funding research and development programs. The following groups participated financially: the agglomerations of Marennes-Oléron and La Rochelle in the Local Action Group Fisheries Aquaculture[12] and the Nouvelle-Aquitaine Regional Council in the Regional Center for Aquaculture Experimentation and Application[13]. The Charente-Maritime department has also developed a coastal water monitoring plan, although water management is outside its statutory competences. Following the identification of pollutants when port dredging – which the department is responsible for – the department has developed a pollution monitoring system to prevent further spreading[14]. It has set up its own coastal water monitoring networks, including: the bathing water quality monitoring network (2011); the shellfish production area quality monitoring network (2014); and the "coastal water" plan (2014). Thus, counting on its own networks (RCD17), the department monitors surface water, adding to the WFD system. It analyzes more than 150 parameters linked to sanitation, agriculture, drinking water supply and swimming. The department also provides technical and financial support to municipalities monitoring the coastal water quality voluntarily, also called active monitoring. Municipalities with bathing areas also benefit from

12 Territorial project funded by the European Maritime Affairs and Fisheries Fund for the sustainable development of local maritime sectors. There are two European Maritime Affairs and Fisheries Funds in this territory.

13 An association whose objective is to promote the development of aquaculture and shellfish farming in the Poitou-Charentes region.

14 The treatment of polluted sludge is expensive.

assistance: "we provide technical assistance for water quality management to municipalities in our department, including coastal territories"[15].

To sum-up, from a geographical point of view, coastal waters are in-between two environmental policies that organize territorial water management. These policies, which aim to achieve good ecological status, are the WFD – for land waters – and the MSFD – for marine environments. In coastal waters, which are first defined by usage (bathing or shellfish farming), both directives apply. The quality of the coastal waters, involves various stakeholders.

2.2. The difficulty of publicizing coastal water quality

The long-standing existence of these policies emphasizes the numerous activities that depend on coastal water quality. Yet, the issue is not at the heart of the political and media agenda. While local press reports on the quality of the beaches every summer, and writes about fishing and marketing bans, the coastal water quality is rarely discussed[16]. In the Pertuis Charentais, the quality of coastal waters and the related consequences, including on shellfish farming, is not a key public matter. Other than a few isolated cases of long-lasting bathing prohibition, coastal water quality in the Charente-Maritime region has not had a negative impact on tourism. Nonetheless, shellfish farmers have been concerned for years about the rise in shellfish mortality. Several mortality events have occurred in hollow oysters (*Magallana gigas*) and common mussels (*Mytilus edulis*) since the 1970s. Between 1969 and 1972, an epizootic disease (gill disease) eradicated the Japanese oyster (*Magallana angulata* or *Crassostrea angulata* depending on the classification), forcing oyster farmers to change their production methods and replace their stock with hollow oysters (*Magallana gigas*). Another critical event was the Oyster Herpes Virus (OsHV-1) epizootic in 2008, which destroyed more than 80% of the hollow oyster spat. Since 2008, abnormal spat and juvenile oyster mortality rates have been observed each year. Since 2012, abnormal adult oyster mortality rates have been noted.

15 Interview with the officer in charge of studying vulnerability profiles of shellfish production areas, coastal water quality units, sea and coastal management, service of the departmental ports of Charente-Maritime (March 8, 2019).

16 We are far from the controversy that the green algae created in Brittany. Refer to [BRU 16] and [LEV 16].

Finally, in 2014 and 2016, mussel mass-mortality rates were noticed. In addition, many shellfish areas have been downgraded, and some faced administrative closures due to poor bacteriological results or the presence of toxic algae in the Pertuis Charentais. To understand and tackle these issues, IFREMER carried out observation and analysis programs (shellfish observation network RESCO in 2009, MYTILOBS in 2012 and MORBLEU between 2015 and 2018). However, even in the case of disasters, an event or a social fact is never a public problem by itself. The agenda-setting of a problem does not depend on the "gravity" of the social problem, but on the political work of policy entrepreneurs (stakeholders, elected representatives, interest groups, etc.)[17]. This political work is even harder when public systems are led by well-established coalitions, as in the case of water policies[18]. It is also impacted by the difficulties surrounding media coverage, the unbalanced powers between public society and farmers, and the almost impossible objectification of coastal water quality.

2.2.1. *The "controlled publicity" of coastal water quality and the consequences on shellfish production in the Pertuis Charentais*

In the Charente-Maritime region, debates on water mainly focus on water quantity[19]. This problem is taken seriously in the Sèvre-Niortaise, Charente and Seudre basins because, since the 1990s, numerous associations have denounced water usage by corn farmers. This means that the controversy over water quantity, its use and management is the result of the mobilization of policy entrepreneurs that succeeded in problematizing and politicizing the issue, by identifying causality between water quantity and agricultural practices and those responsible for such practices. Three arguments can be put forward to explain the difference in the construction of water quantity and water quality problems.

Firstly, coastal water quality impacts coastal stakeholders, who could be both the cause and the first victims of pollution and degradation. Recent

17 The case of the "green algae" problem is emblematic of how public problems are mainly constructed by pointing out the responsibility of a social group (here the agricultural stakeholders), whereas for a long time, agricultural pollution was underestimated and thought as "diffused" [BOU 09].

18 Similar analysis in the field of coastal management [JOR 98].

19 Refer to the extensive literature on water-related conflicts in the Charente basin, for example [GRA 06, MAZ 11, BOU 10].

reflection on plastic waste illustrates this[20]. Roughly 80% of pollution comes from the land, but coastal stakeholders are under-represented within (land) water management commissions. A representative from the commission in charge of the Seudre River Basin stated:

> Officially, the composition should be balanced as the local water commission includes representatives of the agricultural sector, of the shellfish farming, of the environment, of water users. But, in numbers, for one seat, many members from the agricultural area will come in groups, and, although they do not participate each time, we tend to see them more. They dominate. [...] They have a strong influence on writing documents.

The situation is similar in the Charente local water commission, the facilitator in charge of the Charente river basin highlighted:

> The situation is similar in the entire Charente basin. There are bottom-up stakeholders but many more top-down players. The power relationship between the various users exists. There are several factors involved. In addition, shellfish farmers are not as organized. Each department possesses one chamber, which relies on agricultural trade unions and services. It is nationally structured, and implies having lobbyists. I do not mean that the shellfish farmers do not benefit from lobbying. However, the lobbies fight for powers. There are "cartographic, and geographical" power relations, as well as power relations based on the activities.

Farmers quickly understood the need to invest in water management in order to influence policies that directly affect them. Due to their economic weight and organization, they can restrain measures on irrigation, or inputs which are not in their best interests. In addition, they could guide River Basin Management Plans on managing the main issue, namely maintaining water resources to enable the irrigation of field crops, particularly maize. Promoting agricultural interests puts pressure on both water quantity and quality, on which shellfish farming depends. Thus, even if land and shellfish

---

20 Refer to the following example: https://www.sudouest.fr/2019/08/22/ces-dechets-qui-de-rangentca-bouge-dans-le-departement-6469276-1352.php.

farmers share common values that help in maintaining relationships between them [BOS 12], they do not have similar interests or resources. Within local water commissions there does not appear to be a level playing field. Therefore, reports of local water commissions from Charente and Seudre have never mentioned discussions on coastal water quality and shellfish mortality.

Facing a well-structured and dominant coalition, entrepreneurs often resort to media coverage. This strategy, used by environmental groups and fishermen, opened doors for discussions on quantitative water management. The method consisted of showing dead fish through repeated droughts or pointing fingers at corn farmers for using drinking water for irrigation [MAZ 11]. On August 5, 2014, a group of approximately 300 shellfish farmers gathered in La Rochelle to show their anger after mass-mortality hit their oyster farms. They dumped empty shells outside the prefecture to challenge the state and its permissiveness[21]. Professionals condemned "the growth of pesticides in estuaries, the rise in water temperature, the multiplication of pathogens and the dumping of dredging sludge"[22]. However, protests are rare among shellfish farmers. It is difficult for shellfish farmers to point out the responsibility of land farmers, as they share common values [CAN 12]. In addition, developing a media coverage strategy might be tricky. Going public may damage the image of the territory and its activities. The director of the Charente-Maritime regional shellfish farming committee underlined the phenomenon:

> If we mention mortality rates, the marine environment deteriorating, the environmental quality, the message is not that positive. We always aim at making positive statements on the quality of our products, their nutritional interests, their taste, rather than talking about mortality or environmental degradation. But, at some point, remaining passive and not acting will work against us[23].

21 "Oysters, mussels: farmers shout their distress" (Huîtres, moules : les éleveurs crient leur désarroi), *Le Monde*, August 9, 2014.

22 *Idem.*

23 Interview with the Director of the CRC 17 (February 25, 2019).

The third explanation remains in the fact that media coverage is tricky. The pollution is invisible to the naked eye[24], and shellfish farmers cannot rely on expertise to identify causality and responsibility. On the basis of current scientific knowledge, it is difficult to explain the oyster mortality. Despite IFREMER having carried out many studies, the reasons remain poorly understood. Two potential causes were noted: the production methods and the types of species exploited, as well as the poor quality of coastal waters. Nevertheless, no causal connection could be made between the contamination and the mortality rate. With this objective in mind, the Regional Council of Shellfish decided to co-finance a scientific project called AQUAECOs (improvement of the environmental quality and limitation of the pesticides and microplastic toxicity on shellfish activities located in the Pertuis Charentais). The goal is to study the toxicity of microplastics and pesticides on common mussels and hollow oysters, and to then research and provide solutions to reduce toxic impacts.

To sum-up, the degradation of coastal water quality has negative impacts on numerous activities, particularly on shellfish farming, where oysters suffer from excess mortality rates. While the causes of the degradation are unknown, we know for a fact that 80% of pollution in coastal waters comes from the land. However, coastal stakeholders are under-represented in water management bodies and benefit from less influential positions. On the contrary, members of the agricultural sector have significant influence at an economic and political level within the region. Moreover, shellfish farmers cannot implement media strategies that may negatively impact their activities, as they cannot rely on expertise to identify causality and responsibility.

2.2.2. *The quality assessment of variable-geometry coastal waters*

The good status of quantity (such as water flow and groundwater level) is based on the river itself. On the other hand, water quality is measured using standardized indicators[25]. The water quality assessment may easily lead to regular monitoring, thus enabling the classification of areas or bodies of

---

24 This is unlike quantity or nitrate pollution matters, which promote the appearance of green algae.

25 For the development of these indicators, refer to [BOU 18].

water. Based on the WFD, the decree from January 25, 2010 sets biological and physico-chemical parameters to evaluate surface waters. The monitoring control network (RCS) put in place is responsible for carrying out the assessment. Some water bodies benefit from a deadline extension (up to 2012 or 2027), in order to achieve good water status. The monitoring of these areas is dealt with by the operational control network (RCO). The IFREMER then uses the measurements to qualify the status of water bodies. The method detailed in article R212-10 of the Environmental Code is the following: "the status of surface water is determined using the worst measures recorded on both the ecological and chemical statuses." Therefore, a poor result in one of the parameters reduces the overall status of water bodies. A "WFD Atlas", compiling the results, is available for individuals to look at the water quality. They may filter which water body they are interested in[26]. Most of the water bodies located in the Pertuis Charentais are in "good condition".

The MSFD also defines indicators to assess good water status. However, the list is not as stable as that of the WFD, which is an older body. The consistency of the good ecological status of water bodies remains a major issue, mainly for coastal waters. Despite having similar categories of indicators (whether they are physical, biological or hydromorphological), the components studied vary. In addition, for the same descriptor, the preferred indicator may differ. Nonetheless, "scientists setting thresholds and determining good status are for the harmonization of the two directives. This will prevent different regulations from being applied to similar parameters/perimeters"[27]. Thus, to control costs and respect the years of experience of the WFD, the joint WFD–MSFD recovery zone will first harmonize their methods on assessing water status. Therefore, the indicators of good ecological status of marine waters used are those defined by the WFD.

To monitor the biological and chemical parameters of coastal waters, the IFREMER can rely on existing networks that monitor bathing and shellfish waters. However, the status determination of water bodies rests on sectoral

26 This web tool is available using the following: https://wwz.ifremer.fr/envlit/Surveillance-du-littoral/Contaminants-chimiques/Cartographie-Inventaire-du-reseau-ROCCH#/map (accessed 20 May 2021).

27 Interview with a MSFD Project Manager within the "Tropical Networks" group (March 27, 2019).

indicators, which are based on water usage. Public health schemes include monitoring in their plan. Two bacteria, *Escherichia coli* (*E. coli*) and intestinal enterococci, which are known to be dangerous for humans, are particularly investigated. However, the health impact depends on how these bacteria came into contact with humans. Therefore, the thresholds vary according to the prohibition of bathing, fishing or marketing shellfish. In the case of shellfish waters, chemical contaminants are also measured.

Thus, the same water area could be cleared for one use, but not for another. Nothing could prevent a body of water with a good ecological status from being declared of insufficient quality for such activities. Although the articulation of ecological status and chemical status of the WFD plans for a declassification mechanism, there is none in place. Indeed, the poor status of one of the parameters does not reduce the overall water status. Given the socio-economic consequences of limiting or prohibiting uses, the qualitative approach is difficult to adopt. The variable-geometry definition given for coastal water quality blurs the notion of quality, or good status, itself. Indeed, the various classifications based on the measures coexist.

The issue comes more from the abundance, rather than absence, of data available on good ecological status. Yet, current indicators are insufficient to report on the state of coastal waters. The parameters mainly assess the water quality consequences on human health. However, the biological quality of the environment should be considered, as some waters of good quality may prevent the development of species living there. Indeed, the waters in the Pertuis Charentais are in "good condition" and yet the oysters suffer from excess mortality rates. In a CESER report, it was mentioned that "shellfish farmers, by observing their products, benefit from monitoring and defending coastal water quality"[28]. Oysters are sentinel species, "whose behavior, survival and reactions are a 'warning bell' for environmental disruptions"[29]. Although the causes of oysters' death are unknown, mortality is an indicator of insufficient quality assessment. Only a few contaminants, including chemical ones present in sunscreens, and excluding pesticides and micro-plastics, are monitored and measured. The products mentioned that come from human activity impact coastal waters and their effects cannot be measured or regulated.

28 CESER report, op.cit.

29 See: http://www.set-revue.fr/glossaire.

2.3. The Marine Protected Area: going towards a more coherent management of the "Sea–Land" interface?

The recent creation of the Gironde Estuary and Pertuis Marine Protected Area raises hopes among coastal stakeholders, particularly among shellfish farmers. The director of the Regional Shellfish Committee of Charente-Maritime points out: "[the marine park] gives them a voice, no longer preaching in the desert"[30]. The co-financing of the AQUAECOs project gives him hope that the Pertuis Marine Protected Area will become a key player in improving coastal water quality.

> The marine park covers a key territory, within which we are located, and its essence relies on the environmental quality and therefore the marine environment quality. We must take this opportunity. Indeed, it gathers all stakeholders, who could work collectively, discuss challenges and build a management plan or an action plan. The marine park will carry out studies and perform monitoring, as planned in the AQUAECOs project. The marine park provides partial funding to this project[31].

The essence of the Pertuis Marine Protected Area is to protect ecosystems. The marine parks were created in 2006 under the supervision of public state institutions, namely the French National Agency for Biodiversity (AFB)[32]. They have three key objectives: knowledge of the environment, protection of ecosystems and sustainable development of sea-related activities. The Pertuis Marine Protected Area, located in the Gironde estuary, and the Pertuis Sea was created on April 15, 2015. It aims to promote collaborative governance in ecosystems by gathering representatives of local authorities and state services, sea professionals (fishermen, shellfish farmers, port managers, aggregate extractors, etc.), recreational users (boaters, divers, recreational fishermen, sportsmen, etc.), environmental associations, scientific bodies, as well as a management board responsible for defining a territorial plan.

---

30 *Idem.*

31 Interview with the CRC 17 Director, on February 25, 2019.

32 Law No. 2006-436 of April 14, 2006 on national parks, marine natural parks and regional natural parks.

The first management plan dedicated to the Pertuis Marine Protected Area was adopted in June 2018[33]. Thus, it is too early to obverse its impact. As a priority, the Pertuis Marine Protected Area focuses on coastal water quality and preservation of the activities, particularly shellfish farming. To improve the functioning of the marine park ecosystems and protect the water quantity and quality, the management plan mentions the importance of strengthening the "sea–land" connection. This aspect is part of six key goals developed for the Pertuis Marine Protected Area. Nevertheless, the park is not able to regulate land-based activities that negatively impact the water quality and quantity. The management plan briefly mentions principles, such as raising awareness of the consequences of land activities and developing greater "cooperation" with land actors to reach "coordination". While these recommendations are fair, they do not guarantee collaboration between the Pertuis Marine Protected Area and the stakeholders. In addition, the park has no specific expertise in freshwater management. Therefore, even if the National Agency for Biodiversity implies that "the park, one of the largest marine natural parks in mainland France, responds to both natural and human challenges, while considering physical, biological, social and economic components" and develops coherent sea–land management, its capacity to act is limited, including the degradations coming from top-down actors. This could be solved if the Pertuis Marine Protected Area promoted collaboration between coastal stakeholders and strengthened the power relationship with land actors. To date, one cannot affirm this will be the case.

2.4. Conclusion

In 1984, France adopted a coastal law to protect this fragile environment. Although the law originally contained provisions for coastal waters, they are not part of a set of public interventions. On the contrary, their management is particularly fragmented. On one hand, coastal waters, which are interfaces between fresh and marine waters, are an overlapping area between the policies of the WFD and the MSFD. On the other hand, coastal waters are regulated according to their uses (shellfish or bathing waters) and are mainly monitored due to public health reasons. This leads to distortion of the "good condition" or "quality" notions. While quality can be determined using standardized indicators, the preferred indicators rest on the public policy

33 See: https://parc-marin-gironde-pertuis.fr/actualite/approbation-du-plan-de-gestion (accessed 20 May 2021)

objectives and knowledge of contaminants and their effects. Thus, with no measures connecting contaminants to the excess mortality rate observed in oysters, shellfish farmers have few resources to help them raise the issue and discern who is at fault, namely the land actors. Indeed, the waters in the Pertuis Charentais are in "good condition" and yet the oysters are dying. The recent creation of a marine nature park in the Gironde estuary and the Pertuis Sea raises many hopes. Nonetheless, no one can currently tell if this new body will achieve "sea–land" coherence and have the power to act on land activities impacting coastal waters.

2.5. References

[BAR 19] BARONE S., MAYAUX P.-L., *Les politiques de l'eau*, Lextenso Edition, LGDJ, Issy-les-Moulineaux, 2019.

[BOS 12] BOSSUET L., BOUTRY O., "Conflits d'usage et de voisinage autour de la ressource en eau. Illustration à partir du littoral charentais", *Économie rurale*, no. 332, pp. 74–87, 2012.

[BOU 10] BOUBA-OLGA O., BOUTRY O., RIVAUD A., "Conflits d'usage autour de la ressource en eau : une analyse en termes de proximité. Études de cas sur le littoral Poitou-Charentes (France)", in SCHNEIER-MADANES G. (ed.), *L'eau mondialisée. La gouvernance en question*, Éditions La Découverte, Paris, 2010.

[BOU 09] BOURBLANC M., BRIVES H., "La construction du caractère 'diffus' des pollutions agricoles", *Études rurales*, no. 183, pp. 161–176, 2009.

[BOU 18] BOULEAU G., CARTER C., THOMAS A., "Des connaissances aux décisions : la mise en œuvre des directives européennes sur l'eau douce et marine", *Participations*, vol. 21, no. 2, pp. 37–64, 2018.

[BRU 11] BRUN A., LASSERRE F., *Politiques de l'eau : grands principes et réalités locales*, Presses de l'Université du Québec, PUQ, Quebec, Canada, 2011.

[BRU 16] BRUN A., HAGHE J-P., "Les médias, les algues vertes et le modèle agricole breton", *L'Espace géographique*, vol. 45, no. 2, pp. 142–156, 2016.

[CAN 12] CANDAU J., DELDRÈVE V., DEUFFIC P., "Publicisation contrôlée de problèmes territoriaux autour de l'eau. Le cas des Pertuis charentais (France)", *SociologieS*, 2012.

[DRO 13] DROBENKO B., "Les territoires de la GIZC : rénover la vision intégrée de la mer et du littoral. Introduction.", *VertigO – la revue électronique en sciences de l'environnement*, Special edition 18, 2013.

[GAR 10] GARRAUD P., "Agendas/Émergence", in BOUSSAGUET L., JACQUOT S., RAVINET P. (eds), *Dictionnaire des politiques publiques*, Presses de Sciences Po, Paris, 2010.

[GRA 06] GRANJOU C., GARIN P., "Organiser la proximité entre usagers de l'eau : le cas de la Gestion Volumétrique dans le Bassin de la Charente", *Développement durable et territoires. Économie, géographie, politique, droit, sociologie*, no. 7, 2006.

[JOR 98] JORDAN A., GREENAWAY J., "Shifting agendas, changing regulatory structures and the 'new' politics of environmental pollution: British coastal water policy, 1955–1995", *Public Administration*, vol. 76, no. 4, pp. 669–694, 1998.

[LEV 16] LEVAIN A., "Une épreuve médiatique ? Les éleveurs bretons et les marées vertes", *Études rurales*, vol. 198, no. 2, pp. 171–194, 2016.

[MAZ 11] MAZEAUD A., "Quel est le bon débit de l'eau ?", *Pôle Sud*, vol. 35, no. 2, pp. 59–75, 2011.

[PAR 18] PARTICIPATIONS, *La démocratie des territoires de l'eau*, vol. 2, no. 21, Cairn.info, De Boeck Supérieur, Paris, 2018.

3

Comparative Biomarkers Study in Two Scallop Organs to Establish Guidelines for Evaluating French Atlantic Coastline Water Quality

3.1. Introduction

Like many other large ecosystems worldwide, the water of the French Atlantic coastline is subject to diverse anthropogenic pressures (overfishing, climate change, loss of biodiversity and habitats, and chemical pollution because of phytosanitary, organic and inorganic contaminants). Biomonitoring programs follow marine pollution and marine resources from many countries. Because bivalve organisms are exposed to multiple stressors, this study represents research to determine the biomarker response to pollution within the different types of tissue (digestive glands and gills), commonly used in environmental biomonitoring studies [MIL 15, MAI 17, BRE 17]. Indeed, the digestive gland has a detoxification role and the gills are the respiratory organs responsible for dissolving oxygen uptake from seawater [DAL 35].

By definition, a biomarker is "a biochemical, cellular, physiological, or behavioral change that can be measured in body tissues or fluids, or at the level of the whole organism, which reveals the exposure to the effect of one or more chemical pollutants" [DEP 94]. Therefore, some biomarkers are

---

Chapter written by Marine Paomia BREITWIESER, Marine BARBARIN, Valérie HUET, Emmanuel DUBILLOT, Marianne GRABER and Hélène THOMAS.

For a color version of all figures in this book, see www.iste.co.uk/muttin/marine.zip.

currently used for development of the OSPAR Convention for the protection of the Marine Environment of the Northeast Atlantic, such as those for specific metal or organic biological effect monitoring. A battery of biochemical and immunological biomarkers used for pollution assessment were measured in *Mimachlamys varia (M. varia)*, a well-established indicator organism employed in the environmental monitoring programs of the French Atlantic coast [MIL 15, BRE 17, BUS 05, ARI 09, BRE 16]. These biological parameters can offer more relevant information on the potential impact of the chronic chemical contamination of coastal environments. For this reason, and in continuation of previous works on the Atlantic coast, we studied integrative biomarker activity in the different organs of scallops *M. varia*. The black or variegated scallop *M. varia* is a small scallop which lives in benthic areas, and is a species that is representative, in comparison with other bivalves (mussels and oysters), of the environmental quality of semi-closed areas and harbors [BRE 17].

The black scallop *M. varia* is a valuable commercial species in France and attempts have been made to manage natural stocks and replenish those depleted by overfishing [LET 56]. Moreover, we could evoke the notion of sentinel species, which may be used to measure or understand the effects of infectious diseases or air and water contaminants, for example. We note that black scallops occupy a wide geographical range, which is why this scallop species is a sentinel organism for monitoring the health of the oceans [MIL 15, BUS 04]. Finally, a monitoring study using this marine bivalve species (*M. varia*) as a bioindicator for environmental quality assessment presents a new approach for correlating the concentrations of contaminants in the organisms, their health status and genetic diversity [BRE 18].

The aim of this study was (1) to evaluate the potential effect of *in situ* environmental pollution presumed to be in the gills and digestive glands of *M. varia*; (2) to evaluate the sensitivity of both organ types to biomarker modulation.

3.2. Materials and methods

3.2.1. *Study sites and sampling collection*

The study sites are shown in a graphical abstract map. Scallops were collected from 13 sampling sites from Brest to the Nouvelle-Aquitaine

Region, which, to a greater or lesser extent, may be reasonably contaminated areas. Indeed, the biomarkers indicated the state of health in living organisms. Loix on the Île de Ré was the least contaminated site [BUS 05]. Our study identified some measurements of the metallic and organic pollutants in the 13 selected sites (Figure 3.1), and analyzed the relationship between our biomarkers and actual levels of pollution at these sites [BRE 18].

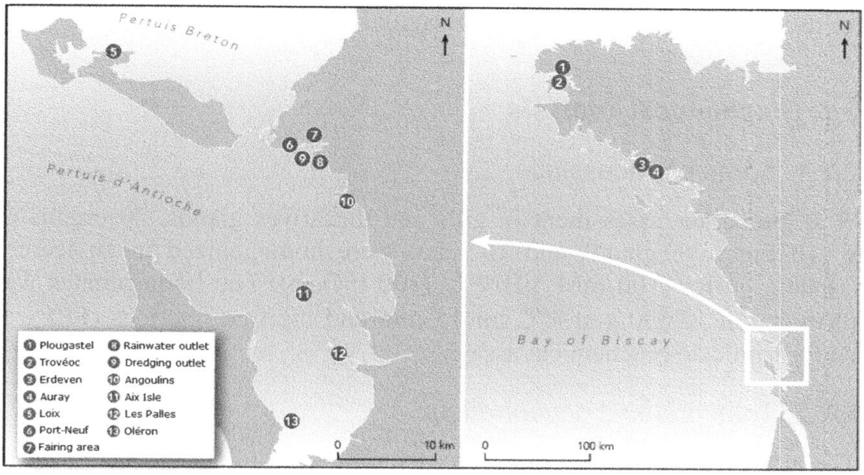

Figure 3.1. *Map of North-West France with the location of the study sites in the Brittany and Nouvelle-Aquitaine regions. Map credit: Pascal Brunello*

In open areas, Troveoc is situated in the Bay of Brest and Erdeven and Auray are subject to contamination from agriculture and intensive farming. Port-Neuf is situated near the wastewater treatment plant and the industrial area of La Rochelle city. Angoulins may be subject to currents and pollution from adjacent urban areas. Aix Isle, Les Palles and Oléron are situated across the Charente estuary, which is impacted by pesticide pollution, and in particular by agriculture (metolachlor, isodrine).

Finally, in semi-open areas, the "dredging outlet" is the area located near tourist beaches and close to the harbor, "rainwater outlet" corresponds to the outlet of rainwater in the harbor (without a water-filtration system) and the "fairing area" is located below the careening area, which has a water-filtration system (carbon, sand, UV and paper to remove organic, inorganic and bacterial contamination).

The digestive glands and gills of 130 variegated scallops (*M. varia*) were collected at each site and placed on dry ice to maintain a temperature of −70°C when traveling to the laboratory (adult size, 10 individuals of each specimen per site for digestive glands and gills).

Once at the laboratory, each of the bivalve organs were stored at −80°C for further analysis. The gills and digestive glands of each specimen were used for the biochemical assessment of biomarkers [BRE 16].

3.2.2. *Biochemical analysis*

3.2.2.1. *Sample processing*

For biomarker assessment in gills and digestives glands, 20 organs per site (10 digestive glands and 10 gills) were homogenized in an ice-cold phosphate buffer (100 mM, pH 7.2, 1100 mOsm). The homogenates were centrifuged at 12,500 g at 4°C for 15 min and the fine fractions (FF) were used for triplicate biochemical assays.

3.2.2.2. *Total protein concentrations*

Total protein concentrations were determined using an adaptation of the BCA Kit methodology (Bicinchononique Acid Kit, Sigma Aldrich). The kit contains bovine serum albumin (BSA) as a standard and involves the reduction of alkaline Cu^{2+} by proteins [SMI 85] with absorbance at 562 nm, using a spectrofluorometer (SAFAS Flx-Xenius).

3.2.2.3. *Antioxidant and detoxification enzyme activity*

SOD activity was assessed in the FF using the method developed by [PAO 86]. The assay, involving EDTA, MnCl2 and mercaptoethanol, measures the decrease in nicotinamide adenine dinucleotide (NADH) oxidation. This inhibition of oxidation was measured at an absorbance of 340 nm (using the SAFAS Flx-Xenius spectrofluorometer) and is a function of SOD activity. Around 50% inhibition of oxidation corresponds to one unit of SOD. The results are presented in U of SOD/mg of protein.

GST activity in digestive glands was determined by adapting a Sigma assay kit method. The Glutathione S-Transferase (GST) assay kit utilizes 1-chloro-2,4-dinitrobenzene (CDNB), which is suitable for the broadest range of GST isozymes. Upon conjugation of the thiol group of glutathione to the CDNB substrate, there was an increase in the absorbance at 340 nm (SAFAS Flx-Xenius). The results are presented in mU of GST/mg of protein [MAN 86, HAB 74].

3.2.2.4. *Laccase assay*

Using a method developed by Luna-Acosta *et al.* [LUN 10b], the activity in the FF was measured. The method is based on the fact that the oxidation of p-phenylenediamine (PPD) is catalyzed by laccase. Thus, the activity of laccase-type phenoloxidase (PO) is a function of PPD degradation. The reaction was assessed at 420 nm using a spectrofluorometer (SAFAS Flx-Xenius). In parallel, the non-enzymatic auto-oxidation (oxidation in the absence of FF) was subtracted. Laccase-type PO activity is expressed as U of laccase/mg of protein (one unit is defined as the amount of enzyme that catalyzes the appearance of 1 μmol of product per min).

3.2.2.5. *Lipid peroxidation*

Damage and, in particular, lipid peroxidation, was estimated by assessing the MDA concentration in the FF [GER 98]. A commercially available MDA assay kit (Oxis International) was used. The method is based on the reaction between MDA and a chromogenic reagent (n-methyl-2-phenylindole). The blue product was quantified by measuring the absorbance at 586 nm [GER 98] using the SAFAS Flx-Xenius spectrofluorometer [MIL 15]. The results are expressed in μM.mg prot^{-1}.

3.2.3. *Data treatments*

Statistical analyses were carried out in R 3.1.2 [RCO 15]. In an initial phase, for all variables, the normality was tested on residuals using Kolmogorov–Smirnov tests and the homogeneity of variances was assessed using Bartlett tests. For biomarkers, the values obtained at the least contaminated site, Loix, were compared between the two organs using a one-way ANOVA.

3.3. Results

Differences across both organs (digestive glands and gills) were apparent for four biomarkers assayed in scallops *M. varia.* Biomarker data from 13 sites is summarized in Table 3.1.

Biomarker		Tissue	Plougastel	Trovéoc	Erdeven	Auray	Loix	Port-Neuf	Rainwater outlet	Dredging outlet	Fairing area	Angoulins	Oléron	Aix Isle	Les Palles
Defense	Superoxide dismutase ($U.min^{-1} mg prot^{-1}$)	Digestive glands	68.8 ± 7 ** a	226 ± 19 *** a	39.2 ± 3.5 a	53.2 ± 4.2 a	25.5 ± 2.5 a	126.0 ± 10 *** a	245.3 ± 14 *** a	21.5 ± 2.3 a	223 ± 17 *** a	281 ± 24 *** a	87.4 ± 11 ***	111.4 ± 8 ***	194.2 ± 12 *** a
		Gills	172.8 ± 15 ***	54.8 ± 5.1 ***	100.3 ± 8.2	119.4 ± 12	126 ± 6	79.2 ± 8.2 ***	85.2 ± 6.9 **	172.1 ± 25 **	68.5 ± 7 ***	137 ± 12	77.7 ± 11 ***	137.0 ± 11	129.2 ± 19
	Glutathion S-transferase ($mU.min^{-1}.mg prot^{-1}$)	Digestive glands	2127 ± 143 ** a	1907.4 ± 88 * a	1818.3 ± 193 a	2493.4 ± 176 ***	1622 ± 145 a	842.9 ± 191 *** a	1036.9 ± 156 *** a	763.6 ± 120 *** a	1736.4 ± 127 a	206.5 ± 58 ***	1103 ± 159 * a	544 ± 85 ***	324.5 ± 80 *** a
		Gills	312 ± 57	165 ± 33	260 ± 39	N/A	403 ± 67	384 ± 16	218 ± 21	103 ± 19 *	342 ± 27	182 ± 27 **	208 ± 30	173 ± 39	138 ± 29 *
Damage	Malondialdehyde ($\mu M.mg prot^{-1}$)	Digestive glands	6.9 ± 0.6 *** a	7.9 ± 0.8 *** a	4.5 ± 0.6	3.6 ± 0.3 a	3.3 ± 0.2	3.9 ± 0.4	3.5 ± 0.4	3.0 ±0.4	4.3 ± 0.5	3.3 ± 1	2.2 ± 0.2	2.7 ± 0.3	4.1 ± 0.6 a
		Gills	3.5 ± 0.3	2.7 ± 0.3	3.7 ± 0.4	7.5 ± 1.9 *	4.7 ± 1.6	3.1 ± 0.3	3.7 ± 0.6	3.7 ± 0.5	2.8 ± 0.4	5.5 ± 0.5	3.6 ± 0.9	3.3 ± 1.2	6.6 ± 1.4
Immuno-modulation	Laccase ($U.min^{-1}.mg prot^{-1}$)	Digestive glands	4.0 ± 0.5 a	6.0 ± 0.5	4.0 ± 0.4	2.9 ± 0.3 a	4.6 ± 0.7	5.9 ± 0.5	6.3 ± 0.7	4.9 ± 0.4	6.3 ± 0.8	8.5 ± 1 ***	7.7 ± 0.8 ***	7.1 ± 0.6 ***	8.2 ± 0.6 ***
		Gills	7.6 ± 0.5 *	7.7 ± 0.9 *	3.9 ± 0.6	6.1 ± 0.9	5.2 ± 0.5	5.8 ± 0.9	7.0 ± 1	4.5 ± 0.4	5.8 ± 1.2	8.4 ± 0.9 ***	6.7 ± 1.2	5.9 ± 2.7	9.1 ± 0.9 ***

Significant differences between Loix (least-contaminated site) and other sites: *$p < 0.05$; **$p < 0.01$; ***$p < 0.001$
Significant differences between digestive glands and gills: a < 0.01
Significant differences between digestive glands and gills: b < 0.05

Table 3.1. *Biomarker responses in digestive glands and gills in* Mimachlamys varia *at the least-contaminated site of Loix (SOD, GST, MDA, Laccase) and study sites at Trovéoc, Plougastel, the fairing area, Rainwater outlet, Angoulins, Les Palles, Port-Neuf, Oléron, Aix Isle, Dredging outlet, Erdeven and Auray. Data is expressed with mean ± SD (n = 10/site with three analyses per individual)*

3.3.1. *Superoxide dismutase*

The mean value of SOD activity in the digestive gland was 25.5 ± 2.5 U/mg in Loix. For this organ, scallops from Plougastel, Trovéoc, the fairing area, the rainwater outlet, Angoulins, Les Palles, Port-Neuf and Aix Isle showed a significantly higher level of SOD activity compared to the four other sites. For the gill sampling, a mean value of 126 ± 6 U/mg of protein (an increase of 83% when compared with the digestive gland) was measured in Loix-en-Ré. Moreover, a lower significant difference was noted between Loix (least contaminated area) and the inside of the harbor areas (fairing area and rainwater outlet), Port-Neuf and Oléron, areas that have been subjected to anthropogenic contamination. On the contrary, Plougastel, and the dredging outlet have higher significant results in comparison with the least contaminated area, Loix.

The activity of this defense enzyme differed significantly in the gills and digestive glands. However, it was observed that there was a specific response from digestive gland organs on zones potentially impacted by local contamination, namely Trovéoc, the fairing area, rainwater outlet, Angoulins, Les Palles, Port-Neuf, Oléron and Aix Isle.

3.3.2. *Glutathione S-transferase*

The mean GST activity in the digestive gland of *M. varia* was 1622 ± 145 mU/mg of protein in Loix-en-Ré. In comparison to this value, the results of the gills of scallops showed a significantly lower level of GST activity for all of the sites ($p < 0.01$). Furthermore, there is a significant positive difference in the enzyme activity of digestive gland organs in Loix (least contaminated site) and three other sites in Brittany (Plougastel, Trovéoc and Auray). On the contrary, in gills, no site has a significant positive difference.

3.3.3. *Malonyldialdehyde*

The Malonyldialdehyde (MDA) content in Loix-en-Ré scallops showed an average value of 3.3 ± 0.2 μM.mg $prot^{-1}$. Two sites (Plougastel and Trovéoc) showed a significant difference to Loix in the MDA content of the digestive gland organ. For the second organ, no significant difference between MDA content was revealed among the sites. Furthermore, the comparison of MDA content in digestive glands shows a 29% increase compared to scallops from Trovéoc, Plougastel and Auray.

3.3.4. *Laccase*

Laccase plays an efficient response role (immunomodulation biomarker) in gills and digestive glands. Indeed, in digestive glands and gills, Laccase activities were 4.6 ± 0.7 and 5.2 ± 0.5 U/mg of protein, respectively, for organisms sampled in Loix (the least contaminated area). When compared to these values, four other sites (Angoulins, Les Palles, Oléron and Aix Isle) presented significantly higher levels of activity for this enzyme in digestive glands and gills. Moreover, a small significant difference was noted between both the digestive glands and the gills for two sites, Auray and Plougastel, which show higher values in gills.

3.4. Discussion

As a result of a 2008 international regulation – the European directive 2008/105/EC on environmental quality standards in the field of water quality – programs are carried out to study and monitor chronic contamination in the different parts of oysters (i.e. the whole body) but rarely in sediment. These very interesting studies in integrator bivalves could, however, have limits, such as whole-body utilization rather than sampling one interesting specialized organ (*Réseau d'Observation de la Contamination CHimique*, ROCCH, 2016).

Here, our research is a directed study on both organs, essential for the evaluation of *in situ* environmental contamination. This approach would allow for the future establishment of a good practice guide to determine key organs in marine environmental quality guidelines, which has never been done before [COT 13, LUN 10a, GUO 17].

The biomarkers of oxidative stress, detoxification and immune modulation were measured in two organs of scallop *M. varia* from 13 sites (in Brittany and Nouvelle-Aquitaine regions). Sampling was carried out in the winter season (low water temperature, normal food availability, no induction of oxygen consumption and no increase of antioxidant defense), which is relevant in biomarker responses in environmental contamination monitoring programs [BRE 18].

Biochemical parameters of oxidative stress, such as SOD, are useful because they can provide the early signs of stress in scallops. The higher SOD activity observed in digestive glands may be linked to the important role of this enzyme in a filter organ in inhibiting ROS (reactive oxygen species) production and thus, decreasing lipid peroxidation effects in cells. This higher activity may constitute an antioxidant response (probably induced by ROS) in *M. varia*. This first finding in *M. varia* is in accordance with previous studies conducted in both organs of small *Chlamys farreri*, which showed a modulation of this enzyme activity due to a cocktail of inorganic and organic contaminants [PAN 09].

The defensive response to contamination was measured in mollusks and may be explained by the Glutathione S-transferase (GST) activity, which plays an important role within the cellular oxidant system and, in particular, within the regulation pathway for the detoxification of xenobiotics. In this study, and in accordance with works in *Chlamys farreri* (three sites) [LIU 12], the activity of GST in the digestive gland was higher ($p < 0.01$) for most sites (12 sites). This could be explained by the different functions of organs. The function of the digestive gland is to serve as the main detoxification mechanism organ and antioxidant synthesis organ, such as glutathione. The gill is in direct contact with water and is thereby more sensitive to the xenobiotic. Therefore, it is deduced that the defense capacity of the digestive gland is higher than in the gill. The difference between tissue compositions may cause the difference in parameter levels, which could explain the MDA results. Indeed, the results highlighted an increased lipid peroxidation in the digestive glands of scallops from two sites (Trovéoc and Plougastel). It was demonstrated that digestive glands function with higher levels of polyunsaturated fatty acids than gills [RIB 91]. The findings of this study could explain the higher levels of lipid peroxidation and detoxification observed in the digestive gland.

Concerning immunomodulation, the values obtained in the digestive glands are in line with previous works in other bivalve species, *Crassostrea gigas* and in scallop *M. varia* [MIL 15, BRE 17, BRE 16, LUN 10a, RIB 91]. Laccase proves to be a sensitive biomarker and phenoloxidase seems to play an important role in bivalve immunity when activated by reactive oxygen species [LUN 10a, RIB 91]. Moreover, Laccase has metal coenzymes that can be affected by heavy metals within a short amount of time, according to the same activation as SOD. This innovative pollution biomarker could be used in both tissues as a biochemical response to environment stressors and could assess the condition of coastal waters. Moreover, since 2015, the Laccase indicator has been involved in the implementation of a normalization tool to monitor the environmental quality in harbor areas [BRE 17] and has permitted the evaluation of chronic chemical contamination in open areas for several years in some mollusks [MIL 15, BRE 17, LUN 10a, RIB 91]. Our results are complementary to the data presented in one of our articles which details a large study on the detection of chemical contaminants and biochemical assays using marine bivalve *Mimachlamys varia* at different sites along the French Atlantic coast [BRE 03].

3.5. Conclusion

In this study, for the first time, we have compared the expression of biochemical and immunological parameters in two organs (gills and digestive glands) in the variegated scallop *Mimachlamys varia*. Our data suggest that the digestive gland may be identified as a key organ for improvements in coastal ecosystem quality. All of our data could be integrated into the standardized procedures to better appraise the status of French Atlantic coastal waters. This data is crucial, not only in obtaining basal information on the health status of *M. varia* in relation to abiotic and biotic factors, but also in relation to activities carried out near the La Rochelle marina, such as the dredging operations necessary for maintaining accessibility and navigational depths, which may have an impact on coastal ecosystems. The ecological relevance of biomarker approaches in the biomonitoring schemes of the French Atlantic coast could propose *M. varia* as a marine sentinel species. Moreover, further research into *M. varia* is needed to strengthen the knowledge base required to establish multivariate mathematical modeling tools and guidelines for evaluating the potential long-term effects of chronic chemical contamination in coastal environments.

3.6. Funding and acknowledgments

M. Breitwieser was a recipient of a Doctoral grant (Communauté d'Agglomération de La Rochelle) and received financial support from the Marina of La Rochelle, in collaboration with the University of La Rochelle and the INCREASE CNRS (Centre National de la Recherche Scientifique) federation to develop this research. The authors would like to thank Michael Cassidy for the English revision. We are deeply indebted to A. Combaud,

T. Guyot, V. Becquet, P. Pineau, N. Lachaussée, A. Viricel from LIENSs (La Rochelle University) and J. Pigeot.

3.7. References

[ARI 09] ARIAS A., FREIRE R., MENDES J. *et al.*, "Intron characterization and their potential as molecular markers for population studies in the scallops *Aequipecten opercularis* and *Mimachlamys varia*", *Hereditas*, vol. 146, no. 2, pp. 46–57, 2009.

[BUS 05] BUSTAMANTE P., MIRAMAND P., "Evaluation of the variegated scallop *Chlamys varia* as a biomonitor of temporal trends of Cd, Cu, and Zn in the field", *Environmental Pollution*, vol. 138, no. 1, pp. 109–120, 2005.

[BRE 16] BREITWIESER M., VIRICEL A., GRABER M. *et al.*, "Short-term and long-term biological effects of chronic chemical contamination on natural populations of a marine bivalve", *PLoS ONE*, vol. 11, no. 3, p. e0150184, 2016.

[BRE 17] BREITWIESER M., VIRICEL A., CHURLAUD C. *et al.*, "First data on three bivalve species exposed to an intra-harbour polymetallic contamination (La Rochelle, France)", *Comparative Biochemistry and Physiology Part C: Toxicology & Pharmacology*, vol. 199C, pp. 28–37, 2017.

[BRE 18] BREITWIESER M., VIGNEAU E., VIRICEL A. *et al.*, "What is the relationship between the bioaccumulation of chemical contaminants in the variegated scallop *Mimachlamys varia* and its health status? A study carried out on the French Atlantic coast using the Path ComDim model", *Science of the Total Environment*, vol. 640–641, pp. 662–670, 2018.

[BUS 04] BUSTAMANTE P., MIRAMAND P., "Interspecific and geographical variations of trace element concentrations in Pectinidae from European waters", *Chemosphere*, vol. 57, no. 10, pp. 1355–1362, 2004.

[COT 13] COTOU E., TSANGARIS C., HENRY M., "Comparative study of biochemical and immunological biomarkers in three marine bivalves exposed at a polluted site", *Environmental Science and Pollution Research*, vol. 20, no. 3, pp. 1812–1822, 2013.

[DAL 35] DALMON J., "Note sur la biologie du pétoncle (*Chlamys varia L.)*", *Revue des travaux de l'Institut des Pêches Maritimes*, vol. 8, no. 3, pp. 268–281, 1935.

[DEP 94] DEPLEDGE M.H., FOSSI M.C., "The role of biomarkers in environmental assessment (2). Invertebrates", *Ecotoxicology*, vol. 3, no. 3, pp. 161–172, 1994.

[GER 98] GERARD-MONNIER D., ERDELMEIER I., CHAUDIERE J. *et al.*, Method of colorimetric analysis of malonic dialdehyde and 4-hydroxy-2-enaldehydes as indexes of lipid per oxidation, kits for carrying out said method, substituted indoles for use in said method and their preparation, US Patent no. US5726063, United States Patent and Trademark Office, Washington, DC, 1998.

[GUO 17] GUO R., PAN L., JI R., "A multi-biomarker approach in scallop *Chlamys farreri* to assess the impact of contaminants in Qingdao coastal area of China", *Ecotoxicology and Environmental Safety*, vol. 142, pp. 399–409, 2017.

[HAB 74] HABIG W.H., PABST M.J., JAKOBY W.B., "Glutathione S-transferases. The first enzymatic step in mercapturic acid formation", *Journal of Biological Chemistry*, vol. 249, no. 22, pp. 7130–7139, 1974.

[LET 56] LETACONNOUX R., AUDOUIN J., "Contribution à l'étude du pétoncle (*Chlamys varia L.*)", *Revue des travaux de l'Institut des Pêches Maritimes*, vol. 20, no. 2, pp. 133–155, 1956.

[LIU 12] LIU N., PAN L., WANG J. *et al.*, "Application of the biomarker responses in scallop (*Chlamys farreri*) to assess metals and PAHs pollution in Jiaozhou Bay, China", *Marine Environmental Research*, vol. 80, pp. 38–45, 2012.

[LUN 10a] LUNA-ACOSTA A., BUSTAMANTE P., GODEFROY J. *et al.*, "Seasonal variation of pollution biomarkers to assess the impact on the health status of juvenile Pacific oysters *Crassostrea gigas* exposed *in situ*", *Environmental Science Pollution Research*, vol. 17, no. 4, pp. 999–1008, 2010.

[LUN 10b] LUNA-ACOSTA A., ROSENFELD E., AMARI M. *et al.*, "First evidence of laccase activity in the Pacific oyster *Crassostrea gigas*", *Fish & Shellfish Immunology*, vol. 28, no. 4, pp. 719–726, 2010.

[MAI 17] MAISANO M., CAPPELLO T., NATALOTTO A. *et al.*, "Effects of petrochemical contamination on caged marine mussels using a multi-biomarker approach: Histological changes, neurotoxicity and hypoxic stress", *Marine Environmental Research*, vol. 128, pp. 114–123, 2017.

[MAN 88] MANNERVIK B., HELENA-DANIELSON U., KETTERER B., "Glutathione transferases – Structure and catalytic activity", *Critical Reviews in Biochemistry and Molecular Biology*, vol. 23, no. 3, pp. 283–337, 1988.

[MIL 15] MILINKOVITCH T., BUSTAMANTE P., HUET V. *et al.*, "*In situ* evaluation of oxidative stress and immunological parameters as ecotoxicological biomarkers in a novel sentinel species (*Mimachlamys varia*)", *Aquatic Toxicology*, vol. 161, pp. 170–175, 2015.

[PAN 09] PAN L., REN J., ZHENG D., "Effect of benzo(a)pyrene exposure on the antioxidant enzyme activity of scallop *Chlamys farreri*", *Chinese Journal of Oceanology and Limnology*, vol. 27, no. 1, pp. 43–53, 2009.

[PAO 86] PAOLETTI F., ALDINUCCI D., MOCALI A. *et al.*, "A sensitive spectrophotometric method for the determination of superoxide dismutase activity in tissue extracts", *Analytical Biochemistry*, vol. 154, no. 2, pp. 536–541, 1986.

[RCO 15] R CORE TEAM, "A language and environment for statistical computing", R Foundation for Statistical Computing, Vienna, Austria, available at: http://www.R-project.org/, 2015.

[RIB 91] RIBERA D., NARBONNE J.F., MICHEL X. *et al.*, "Responses of antioxidants and lipid peroxidation in mussels to oxidative damage exposure", *Comparative Biochemistry and Physiology Part C: Comparative Pharmacology*, vol. 100, nos 1–2, pp. 177–181, 1991.

[SMI 85] SMITH P.K., KROHN R.I., HERMANSON G.T. *et al.*, "Measurement of protein using bicinchoninic acid", *Analytical Biochemistry*, vol. 150, no. 1, pp. 76–85, 1985.

4

Tracer Study and *Mimachlamys varia* Bivalve Tools, a Multidisciplinary Guideline to Assess Environmental Quality in the La Rochelle Harbor Over Three Years

4.1. Introduction

Nowadays, assessing the impact of pollution in coastal waters is key to reaching good chemical and ecological status in coastal environments. Contaminants often lay buried within seafloor sediment but can be released into seas, particularly during cleaning and dredging operations. This phenomenon raises many questions on the impact of such discharges on aquatic ecosystems.

After its expansion in 2014, La Rochelle is one of the largest harbors on the European Atlantic coast, with over 5,000 spaces. Decisive actions and decisions are taken each year in order to comply with environmental commitments, a main feature of the La Rochelle harbor. Indeed, the harbor has been awarded several environmental accreditations, including achieving the ISO 14001 and OHSAS 18001 standards, both of which come

---

Chapter written by Marine Paomia BREITWIESER, Benoit SIMON-BOUHET, Jean-Rémy HUGUET, Valérie HUET, Carine CHURLAUD, Marine BARBARIN, Angélique FONTANAUD and Hélène THOMAS.

For a color version of all figures in this book, see www.iste.co.uk/muttin/marine.zip.

with a list of requirements and improvement targets in terms of environmental performance. These standards afford the harbor "Blue Flag" status, which guarantees good water quality.

Since 2006, ISO 14001 certification has enabled the implementation of sustainable management through various actions available to yachtsmen and amateur sailors. The harbor offers: collection and treatment of cleaning water, following the implementation of a filter system; collection of blackwater and discharged water (from the bilge); selective waste sorting, to collect special waste (waste oils and toxic waste from careening); and the "Blue Brigade" (cleaning up critical waste using special agents).

These actions comply with the EU Water Framework Directive (no. 2000/60/CE) that sets quality goals to assess the chemical and ecological status of coastal waters.

Coastal marine ecosystems are fragile environments that are subject to anthropic (urban development, industry, agriculture, aquatic activities) and environmental pressures. Port discharges are likely to impact upon organisms (fauna and flora). Biological, chemical, chronic and diffuse pollutants affect organism health in port areas.

The following longstanding example perfectly illustrates the contamination in port areas. Antifouling paint used on boats has a high concentration of copper (Cu) to prevent excessive accumulation of marine organisms.

Since 1973, IFREMER (the French Research Institute for Exploitation of the Sea) has monitored concentrations in marine environments, setting up the RNO (French National Observation Network) system. The goal of this network is to monitor the development of contaminant concentrations in marine environments and to assess the quality of coastal waters [CLA 92].

Several research studies focused on improving the quality of port areas using filter feeder organisms such as black scallops *Mimachlamys varia*, collecting dissolved particles [BUS 05, MIL 15]. In this respect, for three years this study aimed to monitor biological and physiological quality and assesses the suitability of the measures taken by the La Rochelle harbor. Therefore, organism samplings were carried out twice a year at six locations, indicated in Figure 4.1 (4 intra-harbor areas: rainwater outlet, fairing area, new fueling station, old fueling station; and 2 extra-harbor sites: dredging outlet and Loix-en-Ré). Biomarkers of effect (e.g. heavy metals) and of exposure (enzymes and metabolites) were then compared to build a standardized, reliable, fast and inexpensive tool dedicated to managers [BRE 17, BRE 18].

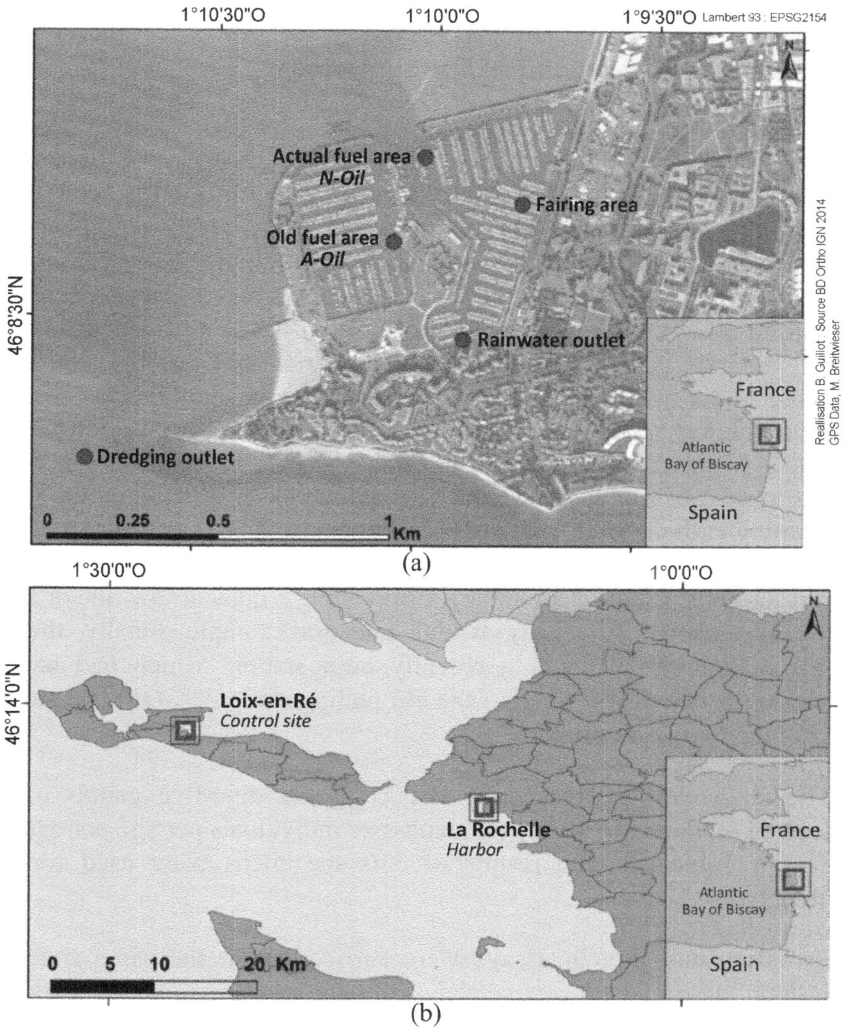

Figure 4.1. *Maps of (a) Les Minimes Marina Harbor and (b) La Rochelle county with the locations of the study areas marked. Loix-en-Ré is the "control site" (as it is the least contaminated) and impacted sites are situated in the La Rochelle Marina: rainwater outlet, fairing area, actual fuel, old fuel and dredging outlet areas*

In addition, a complementary hydrodynamic study of the area was carried out to understand currentology and water mass distribution in the La Rochelle harbor.

4.2. Materials and methods

4.2.1. *Field collection and tissue sampling*

The variegated scallop samples were collected at six locations (Figure 4.1) over three years (2015, 2016, 2017), in winter (January) and in early spring (March). Four of the sites were located inside the La Rochelle harbor (rainwater, fairing, fuels (2); 46°14’50. 15’’N, -1°16’67. 86’’W), one right outside the harbor where mud dredged from the harbor is stocked (dredging outlet) and one site is less contaminated than the others (Loix-en-Ré; 46°23’09.98’’N, -1°41’99.63’’W). Loix-en-Ré is an open area of Ré Island, the “Île de Ré”, chosen as the reference site for this study. The dredging outlet area is near a tourist beach. In the semi-closed port area, rainwater outlets are the zones in which water is drained from sidewalks and freeways (without using a water filtering system). In contrast, the fairing area is located below the careening area, which has a water filtering system (combining carbon, sand, UV and paper filters to remove organic, inorganic and bacterial contamination). The fairing area is the careening and reshaping station for hulls, used for grounding a ship along a quay at low tide or lifting a ship by crane on to a quay at high tide, for example. Finally, the new fueling station (N-Oil) was a recently built station, which had been in operation for two years, whereas the old fueling station (A-Oil) had stopped functioning two years prior.

The study was performed by collecting the digestive glands of 320 variegated scallops (*M. varia*) (10 adult-size individuals per site per date), as shown in Table 4.1. The results of 310 specimens were used for data treatment.

Sampling and protein assays were carried out at the same time. For each sampling site, the digestive glands of bivalves were removed and stored at -80°C for further analysis. The digestive glands were cut into two samples and used to assess trace elements and biomarkers, as previously done in another study [BRE 17, MIL 15]. The samples used for trace element analysis were freeze-dried for 48 hours.

	Site	Contamination characteristics	Specimen number sampled over the three years	Specimen number used for analyses and results
Outside La Rochelle harbor	*Loix-en-Ré*	Control site – reference site	55	52
	Dredging outlet	Salting-out of muds from harbor	55	52
Inside La Rochelle harbor	*Actual fuel area N-Oil*	Recent fueling station for the boats	45	45
	Old fuel area A-Oil	Fuel station that has not been in operation for two years	55	54
	Rainwater outlet	Rain water from public roads	55	54
	Fairing area	Water with fairing, careening residues (after filtered treatment)	55	53
Total			320	310

Table 4.1. *Summary of the design of our field collection of variegated scallops* Mimachlamys varia

4.2.2. *Tracer study of hydrodynamic processes in the harbor area*

A tracer study was conducted in the harbor area in order to characterize the pattern of water flows over the years. The modeling system used was based on the three-dimensional hydrodynamic model TELEMAC-3D [EDF 97], which is part of the TELEMAC-MASCARET system, an integrated modeling tool used to evaluate free-surface flows. TELEMAC-3D solves the Navier–Stokes equations over an unstructured grid and was originally designed to address a large range of spatiotemporal scales (EDF-DRD, 1997). For this study, we used the non-hydrostatic mode to perform simulations at regional (Les Pertuis) and local (the La Rochelle harbor) scales, over a triangular unstructured grid.

Model implementation of the geographic space was discretized using a 40,099 node unstructured grid, the resolution of which ranges from 2.000 m offshore to less than 6 m in the port area of La Rochelle. The distance from the harbor extends to 36 km offshore, 22 km to the north and 77 km to the south. The bathymetry for sub-tidal areas originates from the French Navy (SHOM) and intertidal areas were determined using LIDAR data acquired in 2010. Bathymetry inside the harbor was measured every six months using echo-sounders of the hydrographic service of the Port Atlantique of La Rochelle. Along its open boundaries, the circulation model was forced by 34 tidal constituents (O1, K1, P1, Q1, M2, S2, N2, K2, 2N2, MU2, NU2, L2, T2, M3, M4, MN4, MS4, M6, M8, EPS2, MSF, MSQM, MM, SSA, SA, S4, MKS2, MF, LA2, J1, N4, MTM, R2, S1) obtained by linear interpolation of the global tidal model (FES2014). The modeling system was also forced by hourly fields of sea-level atmospheric pressure and 10 m wind speed (0.2° resolution), data obtained from the National Center for Environmental Prediction (NCEP) and Climate Forecast System Reanalysis (CFSR) [SAH 10]. The hydrodynamic time step was set to 5 s and several simulations (over 2 months) ran on 200 cores, taking 40 h of wall-clock time with a 3D grid using 8 sigma vertical levels (sigma layers).

TELEMAC enables us to consider passive tracers in the model domain. The following equation that describes the evolution of tracer concentration (T) is solved:

$$\frac{\partial T}{\partial t}+U\frac{\partial T}{\partial x}+V\frac{\partial T}{\partial y}+W\frac{\partial T}{\partial z}=\nu\Delta\,(\mathrm{T})+\mathrm{Q} \qquad [4.1]$$

with ν being the tracer diffusion coefficient and Q the source terms for tracers.

4.2.3. *Trace elements and biomarkers of effect analyses*

Analyses of Ag, As, Cd, Co, Cr, Cu, Fe, Mn, Ni, Pb, Se, Sn, V and Zn were performed using a Varian Vista-Pro ICP-OES and a Thermofisher Scientific XSeries 2 ICP-MS. Consequently, aliquots weighing between 60 and 200 mg were digested using a 6:2 (v/v) 67–70% HNO3/34–37% HCl mixture (Fisher, trace metal quality). Acidic digestion of the samples was carried out overnight at room temperature and then in a Milestone microwave oven (30 min at a constantly increasing temperature, up to

120°C, then 15 min at 120°C). Each sample was made of up to 50 mL of ultrapure quality water. For samples weighing less than 100 mg, the mixture used was 3:1 (v/v) 67–70% HNO3/34–37% HCl, and the samples were made of up to 25 mL of ultrapure water. Two certified reference materials (CRMs) and one blank, treated and analyzed in the same way as the samples, were included in each analytical batch. The two CRMs included DOLT-4 (dogfish liver) and TORT-2 (lobster hepatopancreas), both of which came from the National Research Council Canada (NRCC). Mean recovery rates (in %) were of 104 (As), 103 (Cd), 98 (Co), 98 (Cr), 91 (Cu), 88 (Fe), 90 (Mn), 97 (Ni), 92 (Pb), 101 (Se), 95 (V) and 104 (Zn) for TORT2 and 96 (Ag), 92 (As), 97 (Cd), 99 (Cu), 92 (Fe), 101 (Ni), 95 (Pb), 99 (Se) and 106 (Zn) for DOLT-4. Mean element concentrations are expressed in µg/g of dry weight. Quantification limits were of 0.04 (Ag), 0.37 (As), 0.04 (Cd), 0.04 (Co), 0.04(Cr), 0.18 (Cu), 7.38 (Fe), 0.18 (Mn), 0.07 (Ni), 0.04 (Pb), 0.18 (Se), 0.04 (Sn), 0.74 (V) and 7.38 (Zn) µg/g of dry weight [UNR 07].

4.2.4. *Biochemical biomarkers of exposure analysis*

4.2.4.1. *Sample processing*

For biomarker assessment, as described in [BRE 17], samples were homogenized in an ice-cold phosphate buffer (100 mM, pH 7.2, 1100 mOsm). The homogenates were centrifuged at 12,500 g at 4°C for 15 min and the fine fractions (FF) were used for biochemical assays.

4.2.4.2. *Total protein concentrations*

Total protein concentrations were determined using an adaptation of the BCA Kit methodology (Bicinchoninic Acid Kit, Sigma Aldrich). The kit contains bovine serum albumin (BSA) as a standard and involves the reduction of alkaline Cu^{2+} by proteins [SMI 85] with absorbance at 562 nm, using a spectrofluorometer (SAFAS Flx-Xenius).

4.2.4.3. *Antioxidant enzyme activity*

SOD activity was assessed in the FF using the method developed by [PAO 86]. The assay, involving EDTA, MnCl2 and mercaptoethanol, measures the decrease in nicotinamide adenine dinucleotide (NADH) oxidation. This inhibition of oxidation was measured at an absorbance of 340 nm (using the SAFAS Flx-Xenius spectrofluorometer) and is a function of SOD activity. Around 50% inhibition of oxidation corresponds to one unit of SOD. The results are presented in U of SOD/mg of protein.

4.2.4.4. *Detoxification assay*

GST activity in digestive glands was determined by adapting a Sigma assay kit method. The Glutathione S-Transferase (GST) assay kit utilizes 1-chloro-2,4-dinitrobenzene (CDNB), which is suitable for the broadest range of GST isozymes. Upon conjugation of the thiol group of glutathione to the CDNB substrate, there was an increase in the absorbance at 340 nm (SAFAS Flx-Xenius). Results are presented in mU of GST/mg of protein [HAB 74, MAN 86].

4.2.4.5. *Laccase assay*

Using a method developed by Luna-Acosta *et al.* in 2017 [LUN 17], the activity was measured in the FF. The method is based on the fact that oxidation of p-phenylenediamine (PPD) is catalyzed by laccase. Therefore, the activity of laccase-type phenoloxidase (PO) is a function of PPD degradation. The reaction was assessed at 420 nm using a spectrofluorometer (SAFAS Flx-Xenius). In parallel, the non-enzymatic auto-oxidation (oxidation in the absence of FF) was subtracted. Laccase-type PO activity is expressed as U of laccase/mg of protein (one unit is defined as the amount of enzyme that catalyzes the appearance of 1 μmol of product per min).

4.2.4.6. *Lipid peroxidation*

Damage and, in particular, lipid peroxidation, was estimated by assessing the MDA concentration in FF [GER 98]. A commercially available MDA assay kit (Oxis International) was used. The method is based on the reaction between MDA and a chromogenic reagent (n-methyl-2-phenylindole). The blue product is quantified by measuring absorbance at 586 nm [GER 98] using the SAFAS Flx-Xenius spectrofluorometer. The results are expressed in $\mu M.mg\ prot^{-1}$.

4.2.5. *Data analysis*

All statistical analyses were carried out in R 3.4.3 [RCO 15]. Figures 4.2–4.6 were produced in R using the following packages: ggplot2 [WIC 09, pp. 115–137], gridExtra [AUG 16] and factoextra [KAS 17].

The statistical significance of the difference in means between sampling sites or between sampling dates were assessed by Student's t tests when the conditions of normality and homoscedasticity were verified, and otherwise by Wilcoxon tests. The assumption of normality was checked using the Shapiro–Wilk normality test and homoscedasticity using Levene's test for homogeneity of variance across groups from the car package [FOX 11]. Holm's correction was automatically applied in order to consider the increase in the family-wise error rate when performing multiple comparisons.

The principal component analysis was performed using ade4 [DRA 07]. All biomarkers of effect (heavy metal concentrations, 14 variables) and exposure (i.e. specific activity for GST, laccase and SOD and concentration of MDA, 4 variables) were used as variables. In addition, the 297 individuals for whom a complete record was available (i.e. no missing value for any of the 18 variables) were used for observations.

4.3. Results

4.3.1. *Trace elements and biomarkers of effect analyses*

4.3.1.1. *Difference among sites for each year*

Results of trace elements (Pb, Cr, Ni, V, Zn, Cu, Fe) put forward by the EU Water Framework Directive and the OSPAR Commission are shown in Figure 4.2 and sorted by average mean concentrations (all sites). At all six locations, concentrations are expressed in µg/g of dry weight for the following years – 2015, 2016 and 2017 – and averaged over the two seasons (winter and early spring). Contents of the seven metals vary significantly between the intra and extra-harbor sites, and the Loix reference site.

Nickel (Ni) concentrations do not reveal significant variations between sites in 2015 and 2016. Results are substantially homogenous (mean value at Loix in all surveyed years: 2.8 ± 1.2 µg/g). However, in 2017 a significantly lower difference was observed at the old fueling station (A-Oil), compared to the Loix reference site.

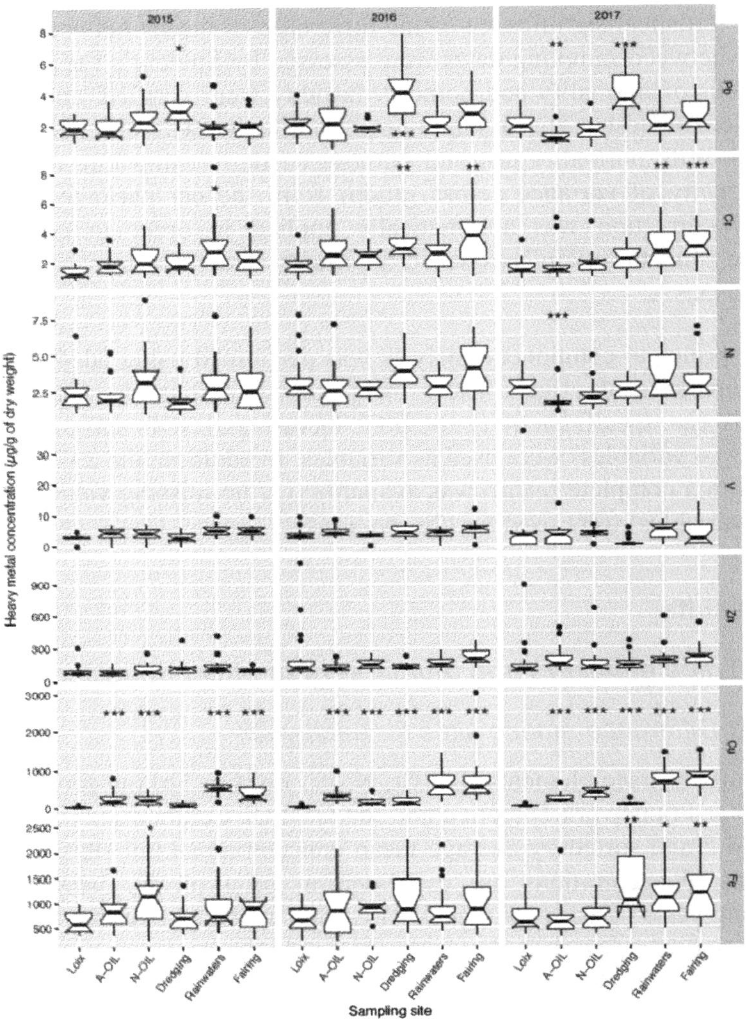

Figure 4.2. *Heavy metal concentrations in the digestive gland of black scallops* M. varia *as a function of sampling site, for each possible combination of heavy metal and sampling year. Boxplot notches are 95% confidence intervals of the median and plain dots are outliers (i.e. points located outside 1.5 times the interquartile range above the upper quartile and below the lower quartile). For each panel, asterisks indicate a significant difference between the mean heavy metal concentration observed in Loix (reference site) and other sampling sites (*: $p < 0.05$, **: $p < 0.01$, ***: $p < 0.001$). Sampling sites are ordered by increasing levels of overall heavy metal contamination (from least contaminated on the left, to most contaminated on the right)*

Vanadium (V) concentrations do not reveal significant variations at potentially impacted sites compared to the Loix site, with mean values of 2.6 ± 1.2 μg/g of dry weight in 2015, 3.4 ± 1.8 μg/g of dry weight in 2016 and 4 ± 8 μg/g of dry weight in 2017.

Zinc (Zn) concentrations presented in Figure 4.2 show numerous extreme points (outliers). Consequently, there is no significant variation between the study sites and the reference site in 2015 (100 ± 55 μg/g of dry weight), 2016 (215 ± 72 μg/g of dry weight) or 2017 (168 ± 68 μg/g of dry weight).

Copper (Cu) concentrations reveal significantly higher differences at all four intra-harbor locations (rainwater outlet, fairing area, old and new fueling stations) in 2015, 2016 and 2017 when compared to the reference site (38 ± 19 μg/g of dry weight in 2015, 33 ± 21 μg/g of dry weight in 2016, 28 ± 20 μg/g of dry weight in 2017). However, in 2017, iron (Fe) concentrations presented significantly higher differences only at the fairing area and the dredging and rainwater outlets, in comparison to Loix (680 ± 254 μg/g of dry weight).

4.3.1.2. *Difference between years for each site*

Figure 4.3 highlights the differences between inorganic contaminants for each study year and site.

Significant variations in lead (Pb) were only noted at the new (N-Oil) and old (A-Oil) fueling stations. A significant variation was observed in 2015/2016 and 2016/2017.

Significant differences in chromium (Cr) were noticed at the old fueling station between 2016 and 2017, with lower values in 2017. The dredging outlet reveals another significant variation between 2015 and 2016. The fairing area showed a significantly higher difference in chromium concentrations in 2016. Cr concentrations remained steady in 2017.

Significantly higher variations in nickel (Ni) were observed at the dredging outlets between 2015 and 2016. In addition, in 2017 vanadium (V) decreased less at the old and new fueling stations than at the dredging outlet.

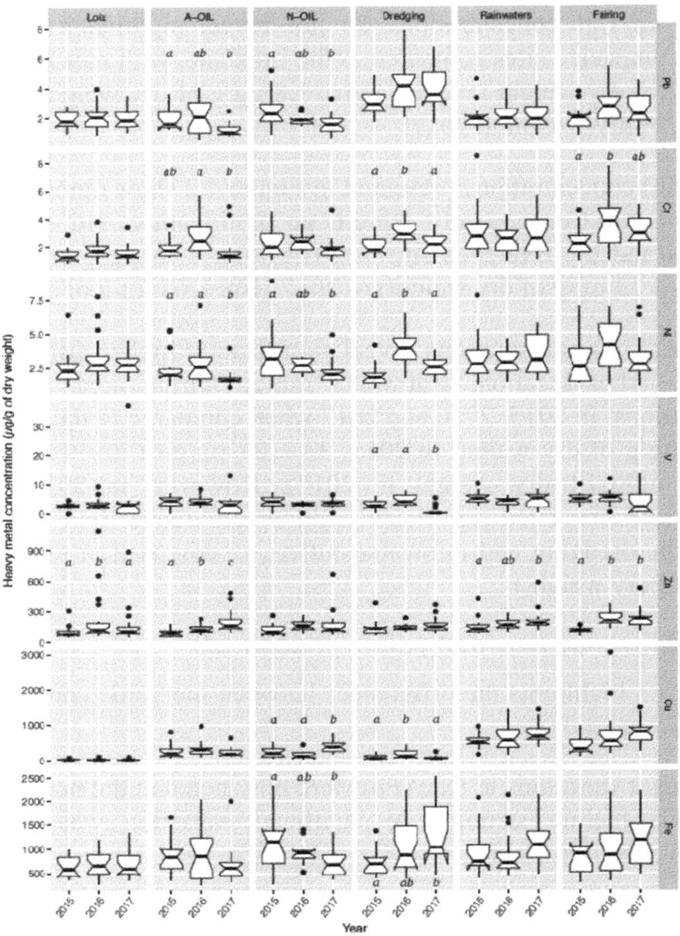

Figure 4.3. *Heavy metal concentrations in the digestive gland of black scallops* M. varia *as a function of sampling year, for each possible combination of heavy metal and sampling site. Boxplot notches are 95% confidence intervals of the median and plain dots are outliers (i.e. points located outside 1.5 times the interquartile range above the upper quartile and below the lower quartile). For each panel, different letters indicate a significant difference between the mean heavy metal concentrations observed in two distinct years. For instance, mean concentration of Pb in Loix is not significantly different in 2015, 2016 and 2017 (no letters, top-left panel). But in A-OIL (next panel on the top row), the mean concentration of Pb is significantly different in 2015 and 2017, while no significant difference was observed between 2015 and 2016 or between 2016 and 2017. Sampling sites are ordered by increasing levels of overall heavy metal contamination (from least contaminated on the left, to most contaminated on the right)*

Zinc (Zn) concentrations show significantly high variations at the reference site, the old fueling station and the fairing area between 2015 and 2016. Moreover, concentrations reveal significantly high variations at the reference site, the fairing area and rainwater outlet, and the old and new fueling stations between 2015 and 2017. In 2017 significantly higher differences were observed at the old fueling station compared to 2016.

Copper (Cu) concentrations reveal significantly higher variations at the dredging outlet between 2015 and 2016, and at the new fueling station (N-Oil) between 2016 and 2017. Results show a decrease in iron (Fe) contents at the new fueling station in 2017.

4.3.2. *Biochemical biomarkers of exposure analysis*

Extracts were taken in order to study oxidative stress response, detoxification and immune capacities, and lipid peroxidation for each study year and site.

4.3.2.1. *Difference between sites for each year*

Figure 4.4 presents the specific activities of SOD, GST, laccase and MDA in the digestive glands.

The GST biomarker is crucial for active protection against polluting or toxic chemical molecules. Profiles revealed higher values of approximately 33% at the old fueling station in 2016. However, results presented great variety in activity. In 2016, lower values of 36% were observed at the new fueling station and the fairing area, when compared with the Loix site (mean value of 1.6 ± 0.2 U/mg of proteins). Results were significantly lower at the fairing area and rainwater outlet in 2017, compared to the Loix site. Mean specific activity of 2.5 ± 0.5 U/mg of proteins was seen at the Loix site in 2017.

The laccase specific activity highlighting immunomodulation was used for the final two years of the study. In 2016 and 2017, the mean values at the reference site were 2 ± 0.3 U/mg of proteins and 6.2 ± 1.2 U/mg of proteins.

In 2016, a significantly higher variation of 30% was noticed at the dredging, careening and rainwater outlets. Nevertheless, in 2017 the dredging outlet showed a downward trend.

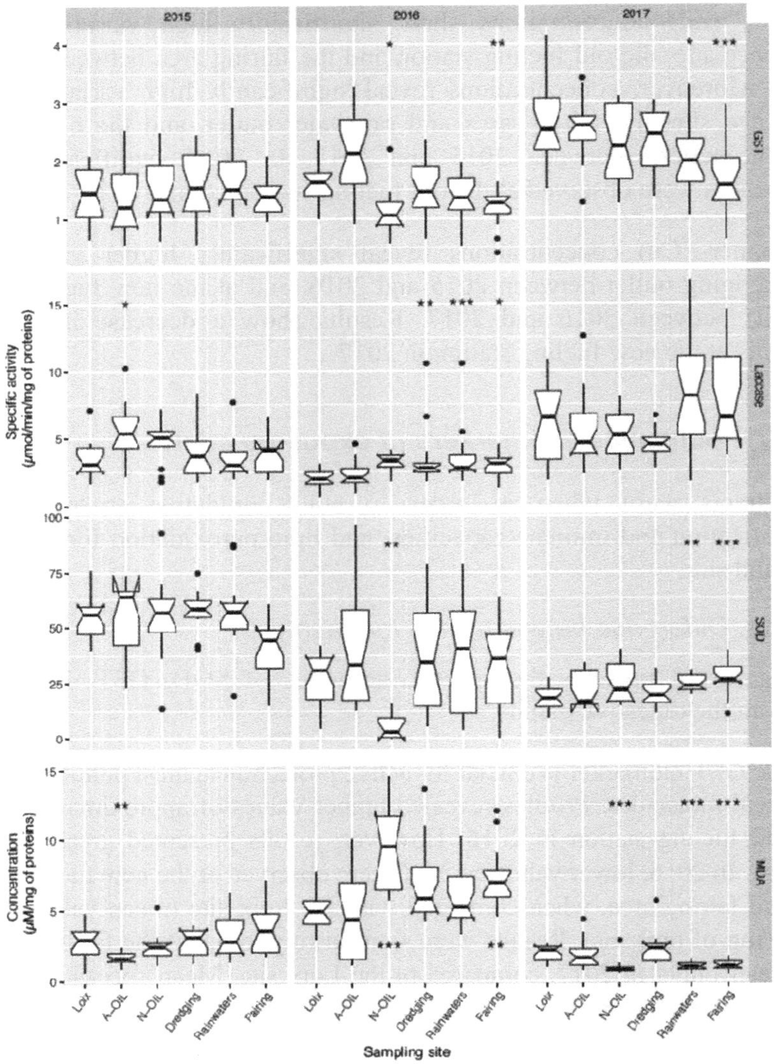

Figure 4.4. *Specific activity (for GST, Laccase and SOD) or concentration (for MDA) of biomarkers assessed in the final fraction of the digestive gland of* M. varia *as a function of sampling site, for each possible combination of biomarker and sampling year. Boxplot notches are 95% confidence intervals of the median and plain dots are outliers (i.e. points located outside 1.5 times the interquartile range above the upper quartile and below the lower quartile). For each panel, asterisks indicate a significant difference between the mean heavy metal concentration observed in Loix (reference site) and other sampling sites (*: p < 0.05, **: p < 0.01, ***: p < 0.001). The ordering of sampling sites is the same as in Figures 4.2 and 4.3*

The SOD biomarker of exposure measured at the reference site reveals mean values of 26.64 ± 4.6 U/mg of proteins in 2016 and 17.63 ± 5 U/mg of proteins in 2017. Compared to the reference site, a significantly lower difference (79%) at the new fueling station in 2016 and a significantly higher variation (25%) at the fairing area and rainwater outlet were observed.

The results of MDA concentrations, a marker of lipid peroxidation, measured in the digestive glands over three years, revealed significant variations at the impacted sites in 2016 and 2017 in comparison to the reference site (5 ± 0.8 µM/mg of proteins in 2016; 2.4 ± 0.6 µM/mg of proteins in 2017). In 2016, a higher difference was observed at the fairing area (+73%) and the new fueling station (+26%). However, in 2017 significantly lower variations were observed at impacted sites, the new fueling station and the fairing area and rainwater outlet (-27%), when compared to the Loix site.

4.3.2.2. *Difference between years for each site*

Figure 4.5 presents the changeability at all six locations between the years and the biomarkers of exposure. The results of the GST specific activities underline higher activity (+53%) in 2017 than in 2015 at the old and new fueling stations, as well as the dredging outlets. They also highlight higher activity (63%) in 2017 than in 2016 at the reference site, fairing area, and the rainwater dredging outlets.

Results of the laccase enzyme revealed lower values of 61% at the reference site, the old (A-Oil) and the new (N-Oil) fueling stations in 2016, when compared to 2015, Moreover, in 2016 and 2017, significant variation (100%) was observed at the reference site, fairing area, the rainwater and dredging outlets and the fueling stations (A-Oil and N-Oil).

Over the three-year period, SOD-specific activity slowly decreased at the Loix site and the old fueling station (A-Oil). For instance, activity decreased by 51% between 2015 and 2016 and by 34% between 2016 and 2017. Furthermore, between 2015 and 2016, SOD-specific activity revealed significantly lower differences at the dredging and rainwater outlets as well as the old and new fueling stations. In contrast, the new fueling station experienced significantly higher specific activity in 2017 than in 2016. Compared to 2015, the first year of the study, lower level signals were observed in 2016. These signals were similar to those at the fairing area and the new fueling station in 2017.

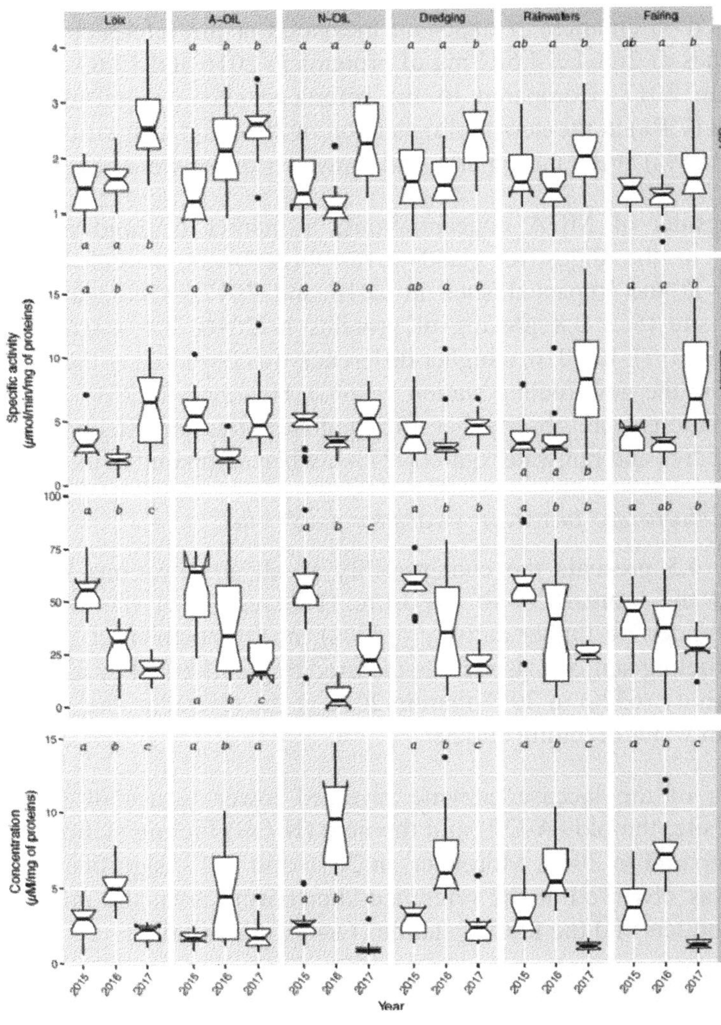

Figure 4.5. *Specific activity (for GST, Laccase and SOD) or concentration (for MDA) of biomarkers assessed in the final fraction of the digestive gland of* M. varia *as a function of year, for each possible combination of biomarker and sampling site. Boxplot notches are 95% confidence intervals of the median and plain dots are outliers (i.e. points located outside 1.5 times the interquartile range above the upper quartile and below the lower quartile). For each panel, different letters indicate a significant difference between the mean heavy metal concentrations observed in two distinct years. For instance, mean specific activity for GST in Loix is not significantly different in 2015 and 2016 (same letter) but a significant difference is detected both between 2015 and 2017, and between 2016 and 2017. The ordering of sampling sites is the same as in Figures 4.2 and 4.3*

The MDA activity revealed significantly higher values of 58% at all six locations between 2015 and 2016. In contrast, a similar decrease was noted at all sites between 2016 and 2017, as well as 2015 and 2017, except at the old fueling station (A-Oil).

4.3.3. *Relationship between biomarkers of exposure (e.g. SOD, GST, laccase, MDA) and effect (e.g. heavy metals)*

Assays of biomarkers of effect (e.g. heavy metals) and exposure (e.g. SOD, GST, laccase, MDA) were performed to assess the influence of contaminants on the health of marine bivalve species. Figure 4.6 presents a descriptive, multidimensional, and easy-to-read analysis that compares results according to site, year and sampling season (January or March).

As shown on the scree plot and the circle of correlation (Figure 4.6), axis 1 has 31.1% of the scatter plot inertia while axis 2 has 12.8%. The circle of correlation enables us to distinguish two color groups: biomarkers of exposure, indicated in blue, (GST, laccase, SOD, MDA), and of effect, in red (heavy metals). Laccase and GST are positively correlated. Both are negatively correlated (180° angle) with MDA, Cd, Pb and Ag. In contrast, MDA and SOD, biomarkers of exposure, and the remaining heavy metals (As, Ni, Co, Cr, Se, V, Fe, Cu, Mn, Sn, Zn) react independently of each other.

Scatter plots represented on the Principal Component Analyses (PCA) revealed:

– the three study years do not present mixed results and are separable due to the variety in overall data;

– in 2017, GST and laccase have a positive influence while MDA, Ag, Cd and Pb have a negative impact;

– under the influence of heavy metals (As, Ni, Co, Cr, Se, V, Fe, Cu, Mn, Sn, Zn), all four intra-harbor locations were differentiated from the 2 extra-harbor sites;

– there is no neat dividing line between sampling seasons (January and March).

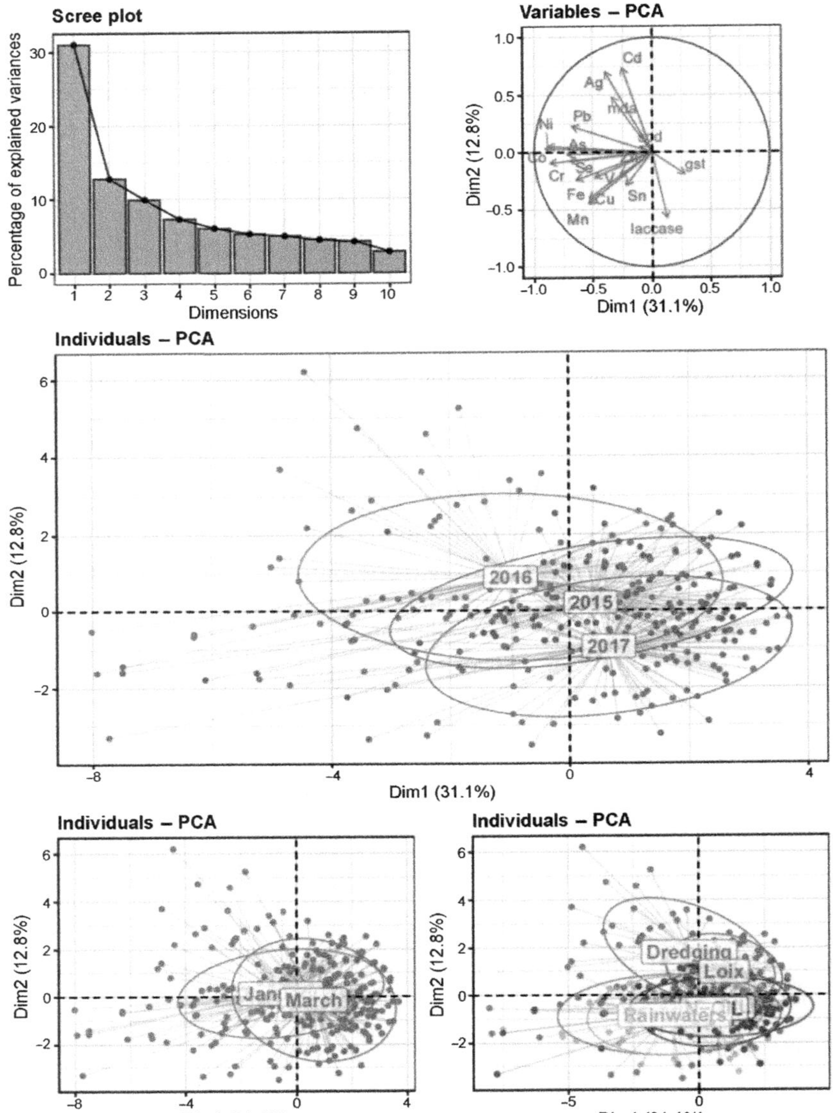

Figure 4.6. *Principal Component Analysis results. Top row: scree plot of eigenvalues for the first 10 factorial axes (left) and correlation circle of the 18 variables used (metals are in red and biomarkers in blue). Middle row: graph of individuals on the first two factorial axes carrying a total of 43.9% of total inertia. Individuals are grouped by sampling year. Bottom row: same as middle row but with individuals grouped by sampling month (left) or sampling site (right)*

4.4. Discussion

The goal of this work was to monitor the quality of the intra-harbor area at La Rochelle for three years. The study looked at (1) identifying the polymetallic chemical contamination of water and (2) assessing the combined response of biomarkers of exposure and effect on black scallops *M. varia* collected at six study sites.

4.4.1. *Bioaccumulation*

Contamination levels at each study site were evaluated by carrying out a study on heavy metal concentration in the digestive glands of the *M. varia* bivalve. This species, located at the water-sediment interface, was carefully chosen as a relevant tool to monitor in the La Rochelle harbor [BRE 17].

According to concentrations in Cr, Ni, Pb, V, Cu, Fe, Zn (Figure 4.2), and in Sn, Co, As, Ag (Figure 4.7), results revealed significantly higher contamination in the port area and the dredging outlets compared to the Loix site, for each year of the study. These trace element values confirm previous work carried out on the same species by [BUS 05, MIL 15, BRE 16, BRE 17]. These concentrations could come from watersheds, for which the port is not responsible.

In recent years, human input has intensified due to the use of various sub-products as phosphate fertilizers – still permitted – metallurgy and vehicle emissions in this semi-closed area.

In addition, variations in metal concentrations (Cn, Zn, Fe) between the three study years were higher at the intra-harbor and dredging sites. In the final two years of the study, these elements accumulated in higher quantities in the digestive glands of the scallops (Figure 4.2). These results could be explained by the handling dynamic which increases over the years (Table 4.2) and a decline in the effectiveness of the filter system at the careening site.

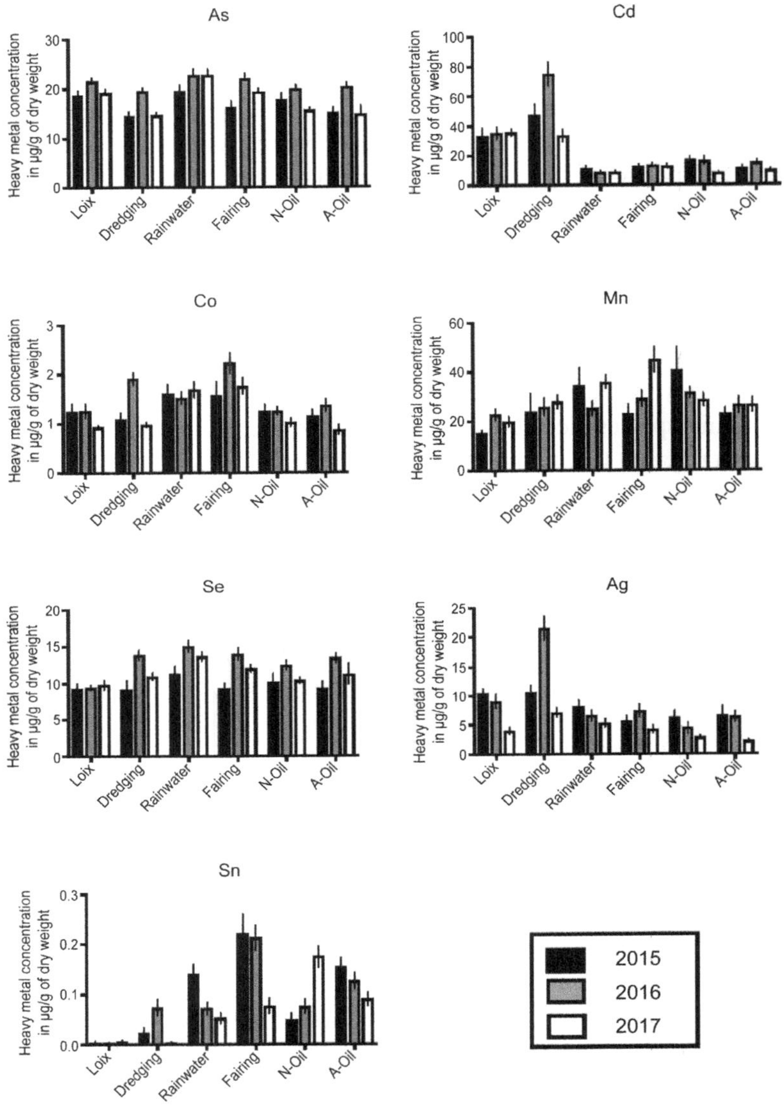

Figure 4.7. *Heavy metals in the digestive gland of black scallops* M. varia *assessed for all sites (Loix, dredging, rainwater, fairing, N-Oil and A-Oil) in 2015, 2016 and 2017 in two seasons (average results of winter and early spring)*

	2015 Boats lifted	2016 Boats lifted	2017 Boats lifted
January	94	117	88
February	345	338.5	364.5
March	414	435	447.5
April	461.5	468.5	427.5
May	389	412.5	438
June	422.5	432.5	429.5
July	311	271	294.5
August	129	152.5	152
September	184	166.5	163.5
October	178.5	0	209.5
November	122	0	151
December	79.5	0	89.5

Table 4.2. *Boats lifted in the fairing area of the La Rochelle harbor in 2015, 2016 and 2017*

Differences in the levels of content could be caused by climatic factors (wind, rain) impacting currentology and thus water temperature. These are key to the bioaccumulation mechanism in bivalve species [HAW 17] and the dilutive effect. For instance, the harbor is a semi-opened area that is connected to the Atlantic Ocean.

It is crucial to take into consideration potential external sources of pollution (e.g. contaminant mixtures – including heavy metals, PCB, PAHs – shown in Table 4.3), which are likely to modify the contaminant levels and impact the water quality in such areas.

		2015		2016		2017		*Tolerance limits* By order of August 9, 2006, December 23, 2009, February 23, 2013, September 17, 2014	
		Rainwater outlet	**Fairing area**	**Rainwater outlet**	**Fairing area**	**Rainwater outlet**	**Fairing area**	***Level 1***	***Level 2***
Heavy metals	**As**	18.0	18.8	16.4	15.3	18.3	17.9	*25*	*50*
	Cd	0.14	0.12	0.12	0.12	0.16	0.12	*1.2*	*2.4*
	Cr	34.3	36.4	36.8	35.6	42.2	42.5	*90*	*180*
	Cu	27.5	34.0	19.4	31.8	22.7	52.2	*45*	*90*
	Ni	19.8	20.8	20.4	19	20.2	21.6	*37*	*74*
	Pb	35.1	33.9	32.4	31.1	35	34.8	*100*	*200*
	Zn	106	113	106	108	110	130	*276*	*252*
	Hg	0.17	0.11	0.12	0.13	0.09	<0.17	*0.4*	*0.8*
PCB	**28**	<0.001	<0.001	<0.001	<0.001	<0.001	<0.001	*0.005*	*0.01*
	52	<0.001	<0.001	<0.001	<0.001	<0.001	<0.001	*0.005*	*0.01*
	101	<0.001	<0.001	<0.001	<0.001	<0.001	<0.001	*0.010*	*0.020*
	118	<0.001	<0.001	<0.001	<0.001	<0.001	<0.001	*0.010*	*0.020*
	138	<0.001	0.0016	<0.001	<0.001	<0.001	<0.001	*0.020*	*0.040*
	153	<0.001	<0.001	<0.001	<0.001	<0.001	<0.001	*0.020*	*0.040*
	180	<0.001	0.0011	<0.001	<0.001	<0.001	<0.001	*0.010*	*0.020*
PAH	**Naphthalene**	0.0100	0.0089	0.0081	0.0150	0.0080	0.0100	*0.1600*	*1.1300*
	Acenaphthylene	<0.0026	<0.0026	0.0031	0.0042	0.0073	0.0072	*0.0400*	*0.3400*
	Acenaphthene	<0.0026	<0.0026	0.0029	0.0050	0.0064	0.0052	*0.0150*	*0.2600*
	Fluorene	0.0250	0.0100	<0.0025	0.0044	0.0064	0.0064	*0.0200*	*0.2800*
	Phenanthrene	0.1100	0.0970	0.0250	0.0380	0.0250	0.0200	*0.2400*	*0.8700*
	Anthracene	0.0100	0.0093	0.0062	0.0095	0.0110	0.0084	*0.0850*	*0.5900*
	Fluoranthene	0.2300	0.1200	0.0440	0.0720	0.0051	0.0330	*0.6000*	*2.8500*
	Pyrene	0.1700	0.0920	0.0380	0.0570	0.0043	0.0290	*0.5000*	*1.5000*
	Chrysene	0.1300	0.0710	0.0200	0.0430	0.0370	0.0220	*0.3800*	*1.5900*

Table 4.3. *Routine measurements of heavy metal concentrations (mg/Kg of dry weight), PolyChloroBiphenyls (mg/Kg of dry weight), and Polycyclic Aromatic Hydrocarbons (mg/Kg of dry weight), in the sediment of two samples from La Rochelle harbor (rainwater outlet and fairing area) for the three years 2015, 2016 and 2017*

4.4.2. *Health*

By performing bivalve sampling at all six locations, this study aimed to evaluate alterations in the biological function (biomarkers of defense) of black scallops, responding to chronic chemical contamination *in situ* [VAN 03]. In 2017, the impacted sites revealed lower activity in lipid peroxidation highlighted by MDA, in comparison to 2015 and 2016.

In 2015 and 2016, at impacted sites (N-Oil, fairing), damage at cell membrane level is coherent with previous work carried out by [LIU 12]. Liu *et al.* studied a similar phenomenon on sites subjected to a contamination gradient of chemical contaminants. The MDA biomarker responds to homeostatic disruption in organisms and could originate from heavy metal intoxication. It is induced by high contamination and would be the consequence of an imbalance between the production of free radicals and defense systems [MIG 11, WIN 91]. The exceeding reactive oxygen species may have two main causes. Firstly, mitochondrial enzyme complexes of the respiratory chain (tetravalent reduction of oxygen) and cells (superoxide radicals) produce an abnormally high quantity which antioxidant products cannot offset. Secondly, the antioxidant system is deficient (SOD, catalase, GPx, GSH) and unable to damage the reactive oxygen species (ROS) produced by the mitochondria [MIG 11]. MDA derives from the oxidation of polyunsaturated fatty acids by the superoxide ion (the main product). The more numerous the superoxide ions are, the higher the MDA quantity a cell has [MIG 11]. The degradation or oxidation of the plasma membrane would be an impact of the contaminants.

The activity of SOD and MDA, the biomarkers of exposure, was evaluated at impacted sites, such as the marina. These enzymes are considered to be early (SOD) and late (MDA) markers of oxidative stress [PAN 05, VAL 06]. For three years and at the majority of the locations, relatively high enzyme activity of SOD attests to a disequilibrium between ROS production and neutralization. For instance, cytosolic SOD requires the existence of metal factors such as copper (Cu), zinc (Zn) and manganese (Mn) for mitochondrial isoforms to function optimally. These heavy metals could be the cause of dysfunctions. Indeed, when these contaminants are spread in sufficient quantity in the water, they may compete with SOD metal factors by replacing them. Therefore, the production of ROS increases rapidly, leading to severe damage in the tissues of marine organisms [AMI 12].

Overall, laccase activity was higher at all sites in 2017, when compared to the other years of the study. This is a sign of strong immune defenses in shellfish. In 2017, the specific activity is higher at the fairing area and the rainwater outlets than at the reference site. In contrast, in comparison to the Loix site, it is lower at the dredging outlet.

This negative immunomodulation at the dredging outlet could be explained by the impact of inorganic and organic contaminants (hydrocarbons and agrochemicals) on laccase. This enzyme is particularly sensitive to these contaminants [LUN 10, LUN 17]. In the case of pollution by contaminant mixtures, the organism experiences an increase in ROS which may lead to overall enzyme overactivation, along with intensification of the specific activity response of the laccase immunomarker [LUN 17].

In 2016, the phase II enzyme GST (bioinactivation, conjugation) acting in the detoxification mechanisms of xenobiotics was higher at the old fueling station, compared to the Loix site. This antioxidant enzyme would be particularly active at the spot where xenobiotics are in suspension in the water column or stuck in the mud. Thus, the black scallop *M. varia*, which has a suprabenthic situation close to mudflats, would have its enzymes (GST) involved in detoxification and antioxidant protection of activated organic components [ALD 99]. This conclusion is similar to two studies led by [PAN 05] that highlighted the increase in GST activity in the digestive glands of mollusks (*C. farreri*) due to moderate concentrations of pesticides and PAHs.

Heavy metals partly play a role in the enzyme response involved in the antioxidant and detoxification activities of the biomarker of exposure MDA, which is positively correlated with Pb, Ag and Cd. However, these three components are negatively correlated with SOD, laccase and GST.

These different correlations suggest that oxidative stress may be linked to (1) heavy metals from bioaccumulation and/or (2) contamination of anthropized waters. Results are coherent with previous studies on the same species and at the same sampling locations [BRE 16, MIL 15]. They also underline negative correlations between GST and laccase and Cd, Ag and MDA. This could explain the loss of effectiveness of the Warburg–Dickens–Horecker shunt of these two biomarkers. The study revealed correlations between heavy metal concentrations and the specific enzyme activity of biomarkers of exposure. However, other *in situ* parameters could impact this response. Several natural abiotic factors including oxygen level, temperature and salinity can significantly modify the oxidative stress of aquatic animals [LUS 11]. In addition, they can influence the immunological parameters that are visible through the laccase assay [WAN 12].

4.4.3. *Tracer study in the bay of La Rochelle*

3D modeling of the hydrodynamic processes and tracer studies performed in the La Rochelle harbor enabled us to distinguish several patterns that are specific to the tidal and meteorological phenomena of the area. A video [BRE 19] (screenshots of February 7, 2016, given in Figure 4.10) shows the journey of a tracer, consistently released at the careening station in modeled hydrodynamic conditions between January 30 and February 16, 2016. All tidal and meteorological conditions have been shown to visualize the effect of different forcings.

4.4.3.1. *Role of the tide*

Overall, from rising mid to high tide, the tracer remained in the port area in the northern parts of the basin, particularly in the basin of origin (between the fairing area and the rainwater outlets – shown in Figure 4.1). When the tide went out, the tracer headed towards the channel to pour out of the harbor and spread rapidly across the bay. The tracer released in the bay tends to go south at low tide before reaching the beaches of southern La Rochelle (Châtelaillon) at high tide.

At spring tide, the currents generated by the tide are very fast, frequently overpassing 1.5 m.s-1 at the channel entrance of the port. From low tide to high tide this intense flow, coming from offshore, enters the port via the channel, which generates eddies at the scale of each of the sub-basins. During ebb tide, the tracer was more confined to the basin into which it was released and was drained by the channel entrance. At slack tides the tracer remained totally confined to its sub-basin, only spreading by diffusion. Finally, the tracer diffusion to other basins was lower during neap tides, where meteorology can have a bigger influence because water fluxes linked with the tide are less intense. The eddies created by the advective flow into the port are actually weaker at neap tide than at spring tide, and water masses can be moved more easily.

4.4.3.2. *Role of the meteorology*

Various winds prevail in the Pertuis Charentais: in order of occurrence, westerly, north-westerly, north-easterly and southerly winds are the most common, even if the area is subject to a heterogeneous forcing throughout the year. Their impact on the port, navigation and bay traffic can be considerable. Overall, qualitative analysis has shown that southerly winds tend to move the tracer northward, that is towards the Pertuis Breton, through the narrow passage between the Île de Ré and La Rochelle. Strong westerly or even north-westerly winds, frequent in winter during storms, diffused the tracer into the entire bay of La Rochelle, causing its journey to move further south. Strong north-easterly winds pushed the tracer to the south while accelerating its diffusion throughout the entire harbor.

The macro-tidal regime of La Rochelle gives the area a strong dynamic, capable of diffusing a pollutant that is released in the harbor to the open sea. Only one tide is necessary to diffuse the tracer outside the port, whether a neap or spring tide. As tides have been globally homogeneous in the area over the years, they cannot be used as an indicator to compare toxicity levels from one year to another. Only an irregular forcing, such as meteorology, could explain the different results obtained.

Comparisons with atmospheric forcing between 2015–2016 and 2016–2017 revealed inter-annual variability of the direction of the prevailing wind. This variability was shown by [DOD 10] to be partly controlled by the North Atlantic Oscillation.

The compass rose (Figure 4.8) shows that the strongest winds from March 2015 to January 2016 came from the west and, to a lesser extent, the north-west, whereas north-easterly and southerly winds had a less significant influence. Figure 4.9 highlights the strong influence of north-easterly winds from April 2016 to January 2017. As for the preceding year, we can see that westerly and north-westerly winds prevail.

As shown in the aforementioned video [BRE 19], strong north-easterly winds encourage water bodies to leave the bay of La Rochelle. These winds, which were very powerful during the April 2016–January 2017 period favor the redistribution and diffusion of water bodies in the harbor. In comparison, the period between March 2015 and January 2016, considered a "normal" period in terms of the wind forcing, tended to diffuse and redistribute the water masses in the bay less. These results are coherent with *in situ* observations, which indicate higher toxicity in 2016 than in 2017. This could be linked to a lower redistribution of the water bodies in 2015–2016 than in 2016–2017.

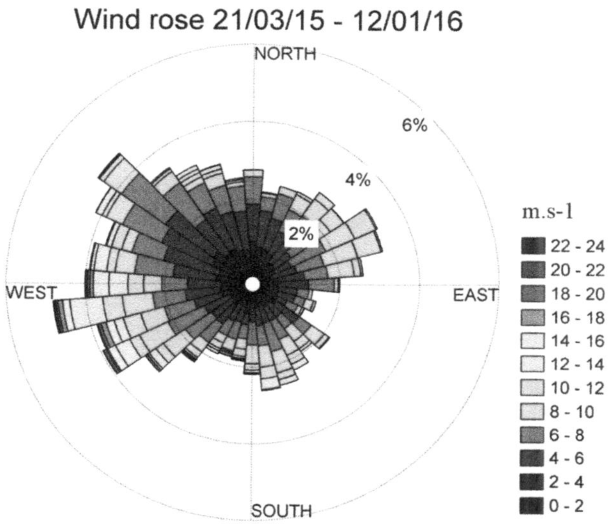

Figure 4.8. *Wind rose between the years 2015 and 2016*

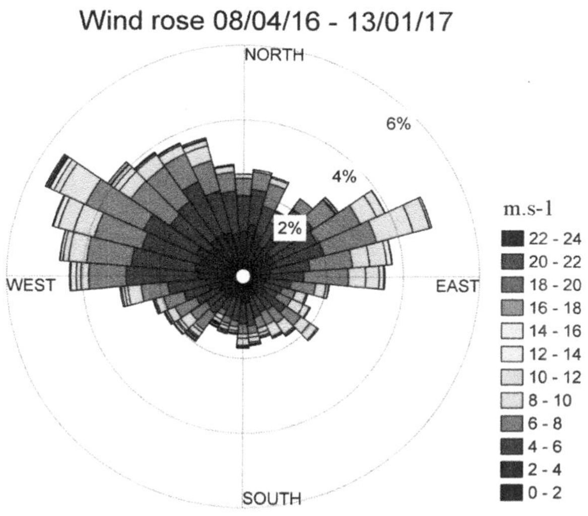

Figure 4.9. *Wind rose between the years 2016 and 2017*

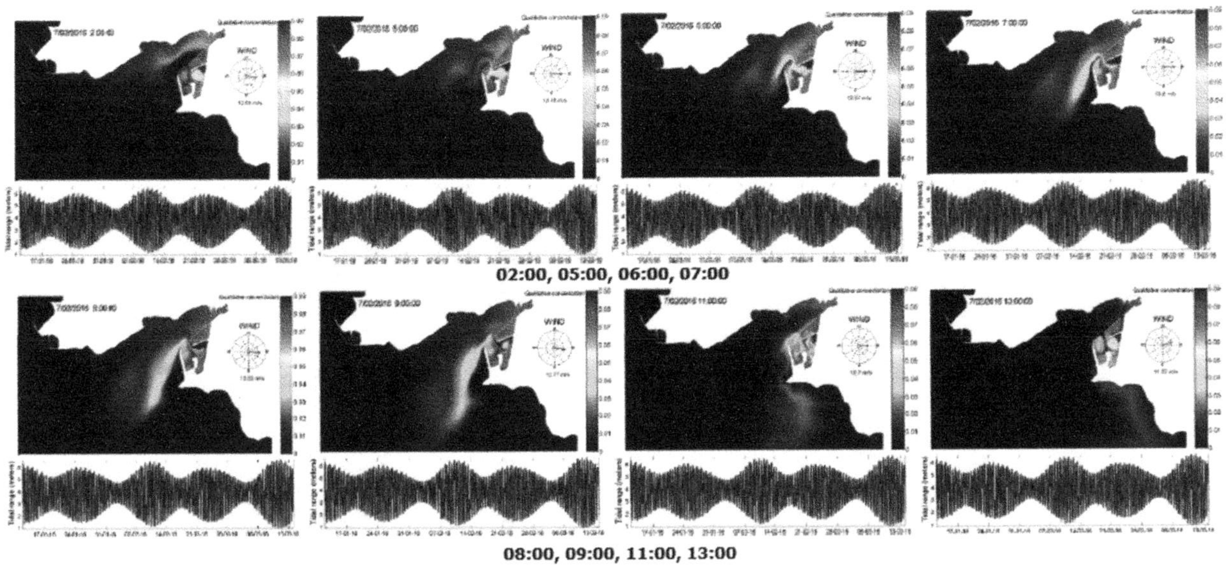

Figure 4.10. *Screenshots of the video [BRE 19]. Hydrodynamic modeling approach carried out in the La Rochelle Marina to illustrate the qualitative concentration of particles in the function of a wide and tidal range. February 7, 2016 at 02:00, 05:00, 06:00, 07:00 and 08:00, 09:00, 11:00, 13:00*

4.5. Conclusion

After being awarded the ISO 14001 certification, the La Rochelle harbor has contributed to environmental protection and the environmental awareness of its users. The harbor authority currently works to reduce the level of pollution. With this aim in mind, in 2014, the harbor staff implemented a system of careening water treatment, as well as a new fueling station (N-Oil) with an oil skimmer and hydrocarbon separator.

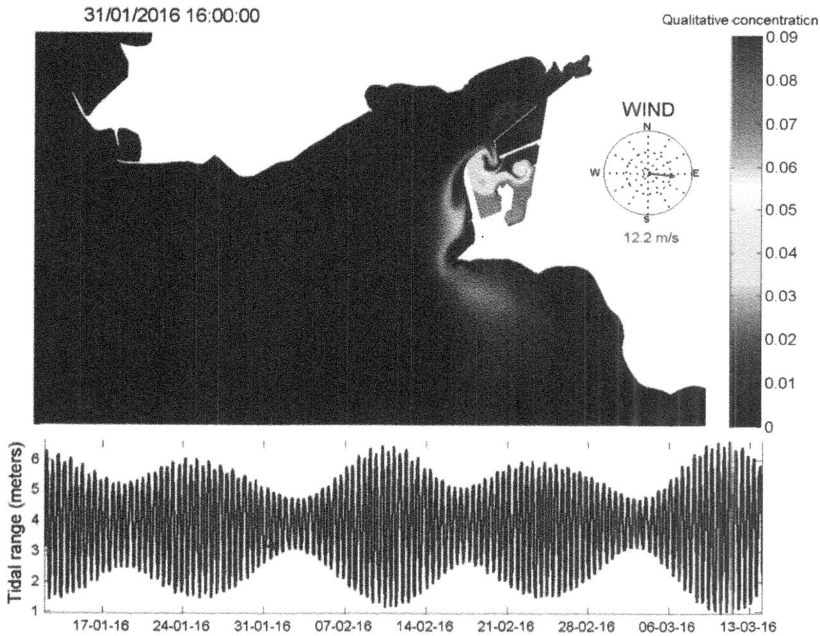

Figure 4.11. *Hydrodynamic modeling approach carried out in the La Rochelle Marina to illustrate qualitative concentration of particles in function of wide and tidal range. Video extract, see [BRE 19]*

This study focused on monitoring the water quality to assess the effectiveness of the measures and facilities implemented and was carried out at different periods over three years. It was also essential to consider sources of potential external pollution which would likely modify the contaminant level in a semi-closed space. Meteorological characteristics, estuaries and the treatment plant located near the La Rochelle harbor may introduce chemical and biological pollutants through the various inter-annual atmospheric forcings (video extract [BRE 19] shown in Figure 4.11). Hydrodynamics can affect water quality in the harbor partially or significantly as the currents and water flows in the area are subject to high intensity meteorological and tidal forcings.

Results of this study suggest that using a multi-biomarker approach on black scallops *Mimachlamys varia* could be an effective method to respond to scientific interrogations in ecotoxicology. This method could be used to better understand human pressures on intra-harbor marine ecosystems. In addition, this innovative work, performed *in situ* for such a long period, revealed data that has never been presented before.

4.6. Funding and acknowledgments

M. Breitwieser was a recipient of a Doctoral grant (Communauté d'Agglomération de La Rochelle) and received financial support from the Marina of La Rochelle, in collaboration with the University of La Rochelle, to develop this research. The authors would like to thank the people involved in sampling: A. Combaud, T. Guyot, V. Becquet, Emmanuel Dubillot. We are also grateful to Célia Fery and Bastien Masson for performing some biomarker assays.

4.7. References

[ALD 99] ALDER V., YIN Z., TEW K.D. *et al.*, "Role of redox potential and reactive oxygen species in stress signalling", *Oncogen*, vol. 18, no. 45, pp. 6104–6111, 1999.

[AMI 12] AMIARD-TRIQUET C., AMIARD J.-C., RAINBOW P.S., *Ecological Biomarkers: Indicators of Ecotoxicological Effects*, CRC Press, Boca Raton, FL, 2012.

[AUG 16] AUGUIE B., "gridExtra: Miscellaneous functions for 'grid' graphics", R package version, no. 2(1), p. 39, available at: https://CRAN.R-project.org/package=gridExtra, 2016.

[BRE 16] BREITWIESER M., VIRICEL A., GRABER M. *et al.*, "Short-term and long-term biological effects of chronic chemical contamination on natural populations of a marine bivalve", *PLoS ONE*, vol. 11, no. 3, p. e0150184, 2016.

[BRE 17] BREITWIESER M., VIRICEL A., CHURLAUD C. *et al.*, "First data on three bivalve species exposed to an intra-harbour polymetallic contamination (La Rochelle, France)", *Comparative Biochemistry and Physiology Part C: Toxicology & Pharmacology*, vol. 199C, pp. 28–37, 2017.

[BRE 18] BREITWIESER M., VIGNEAU E., VIRICEL A. *et al.*, "What is the relationship between the bioaccumulation of chemical contaminants in the variegated scallop *Mimachlamys varia* and its health status? A study carried out on the French Atlantic coast using the Path ComDim model", *Science of the Total Environment*, vols 640–641, pp. 662–670, 2018.

[BRE 19] BREITWIESER M., THOMAS-GUYON H., MUTTIN F., "Hydrodynamic modelling of La Rochelle Marina harbour from 2016 January 30th to February 16th with tracer continuous release at careening station, LIENSs, CNRS, La Rochelle University", *MédiHAL*, mp4 video, available at https://medihal.archives-ouvertes.fr/hal-03038687 [accessed December 2020], 2019.

[BUS 05] BUSTAMANTE P., MIRAMAND P., "Evaluation of the variegated scallop *Chlamys varia* as a biomonitor of temporal trends of Cd, Cu, and Zn in the field", *Environmental Pollution*, vol. 138, no. 1, pp. 109–120, 2005.

[CLA 92] CLAISSE D., JOANNY M., QUINTIN J.-Y., "Le réseau national d'observation de la qualité du milieu marin (RNO) ", *Analusis*, vol. 20, no. 6, pp. M19–M22, 1992.

[DOD 10] DODET G., BERTIN X., TABORDA R., "Wave climate variability in the North-East Atlantic Ocean over the last six decades", *Ocean Modelling*, vol. 31, nos 3–4, pp. 120–131, 2010.

[DRA 07] DRAY S., DUFOUR A.-B., "The ade4 package: Implementing the duality diagram for ecologists", *Journal of Statistical Software*, vol. 22, no. 4, pp. 1–20, 2007.

[EDF 97] EDF-DRD, TELEMAC modelling system TELEMAC-3D, Version 2.2 – Système de modélisation TELEMAC-3D, Note Théorique, no. HE-42/97/049/B, 1997.

[FOX 11] FOX J., WEISBERG S., *An R Companion to Applied Regression*, Sage Publications, Los Angeles, 2011.

[GER 98] GERARD-MONNIER D., ERDELMEIER I., CHAUDIERE J. *et al.*, Method of colorimetric analysis of malonic dialdehyde and 4-hydroxy-2-enaldehydes as indexes of lipid per- oxidation, kits for carrying out said method, substituted indoles for use in said method and their preparation, US Patent no. US5726063, United States Patent and Trademark Office, Washington, DC, 1998.

[HAB 74] HABIG W.H., PABST M.J., JAKOBY W.B., "Glutathione S-transferases. The first enzymatic step in mercapturic acid formation", *Journal of Biological Chemistry*, vol. 249, no. 22, pp. 7130–7139, 1974.

[HAW 17] HAWKINS C.A., SOKOLOVA I.M., "Effects of elevated CO_2 levels on subcellular distribution of trace metals (Cd and Cu) in marine bivalves", *Aquatic Toxicology*, vol. 192, pp. 251–264, 2017.

[KAS 17] KASSAMBARA A., MUNDT F., factoextra: Extract and Visualize the Results of Multivariate Data Analyses, R package version 1, 2017.

[LIU 12] LIU N., PAN L., WANG J. *et al.*, "Application of the biomarker responses in scallop (*Chlamys farreri*) to assess metals and PAHs pollution in Jiaozhou Bay, China", *Marine Environmental Research*, vol. 80, pp. 38–45, 2012.

[LUN 10] LUNA-ACOSTA A., ROSENFELD E., AMARI M. *et al.*, "First evidence of laccase activity in the Pacific oyster *Crassostrea gigas*", *Fish and Shellfish Immunology*, vol. 28, no. 4, pp. 719–726, 2010.

[LUN 17] LUNA-ACOSTA A., BREITWIESER M., RENAULT T. *et al.*, "Recent findings on phenoloxidases in bivalves", *Marine Pollution Bulletin*, vol. 122, nos 1–2, pp. 5–16, 2017.

[LUS 11] LUSHCHAK V., "Environmentally induced oxidative stress in aquatic animals", *Aquatic Toxicology*, vol. 101, no. 1, pp. 13–30, 2011.

[MAN 88] MANNERVIK B., HELENA-DANIELSON U., KETTERER B., "Glutathione transferases – Structure and catalytic activity", *Critical Reviews in Biochemistry and Molecular Biology*, vol. 23, no. 3, pp. 283–337, 1988.

[MIG 11] MIGDAL C., SERRES M., "Espèces réactives de l'oxygène et stress oxydant", *Médecine/Sciences*, vol. 27, no. 4, pp. 405–412, 2011.

[MIL 15] MILINKOVITCH T., BUSTAMANTE P., HUET V. *et al.*, "*In situ* evaluation of oxidative stress and immunological parameters as ecotoxicological biomarkers in a novel sentinel species (*Mimachlamys varia*)", *Aquatic Toxicology*, vol. 161, pp. 170–175, 2015.

[PAN 05] PAN L., REN J., LIU J., "Effects of benzo (k) fluoranthene exposure on the biomarkers of scallop *Chlamys farreri*", *Comparative Biochemistry and Physiology Part C: Toxicology & Pharmacology*, vol. 141, no. 3, pp. 248–256, 2005.

[PAO 86] PAOLETTI F., ALDINUCCI D., MOCALI A. *et al.*, "A sensitive spectrophotometric method for the determination of superoxide dismutase activity in tissue extracts", *Analytical Biochemistry*, vol. 154, no. 2, pp. 536–541, 1986.

[RCO 15] R CORE TEAM, A language and environment for statistical computing, R Foundation for Statistical Computing, Vienna, Austria, available at: http://www.R-project.org/, 2015.

[SAH 10] SAHA S., MOORTHI S., PAN H.L. *et al.*, NCEP climate forecast system reanalysis (CFSR) 6-hourly products, January 1979 to December 2010, Research Data Archive at the National Center for Atmospheric Research, Computational and Information Systems Laboratory, Boulder, USA, 2010.

[SMI 85] SMITH P.K., KROHN R.I., HERMANSON G.T. *et al.*, "Measurement of protein using bicinchoninic acid", *Analytical Biochemistry*, vol. 150, no. 1, pp. 76–85, 1985.

[UNR 07] UNRINE J.M., HOPKINS W.A., ROMANEK C.S. *et al.*, "Bioaccumulation of trace elements in omnivorous amphibian larvae: Implications for amphibian health and contaminant transport", *Environmental Pollution*, vol. 149, no. 2, pp. 182–192, 2007.

[VAL 06] VALAVANIDIS A., VLAHOGIANNI T., DASSENAKIS M. *et al.*, "Molecular biomarkers of oxidative stress in aquatic organisms in relation to toxic environmental pollutants", *Ecotoxicology and Environmental Safety*, vol. 64, no. 2, pp. 178–189, 2006.

[VAN 03] VAN DER OOST R., BEYER J., VERMEULEN N.P.E., "Fish bioaccumulation and biomarkers in environmental risk assessment: A review", *Environmental Toxicology and Pharmacology*, vol. 13, no. 2, pp. 57–149, 2003.

[WAN 12] WANG Y., HU M., CHEUNG S.G. *et al.*, "Immune parameter changes of hemocytes in green-lipped mussel *Perna viridis* exposure to hypoxia and hyposalinity", *Aquaculture*, vol. 356, pp. 22–29, 2012.

[WIC 09] WICKHAM H., *ggplot2: Elegant Graphics for Data Analysis*, Springer-Verlag, New York, 2009.

[WIN 91] WINSTON G.W., DI GIULIO R.T., "Prooxidant and antioxidant mechanisms in aquatic organisms", *Aquatic Toxicology*, vol. 19, no. 2, pp. 137–161, 1991.

5

ISDAMP Project Dissemination and Continuation for Local Authorities' Oil Spill Preparedness

Marine pollution research projects often seek to extend their results to other regions. Rather than addressing this type of project extension, we focus on the different categories of stakeholders who were interested in the ISDAMP project. Our actions were shown to international and local authorities so that they could participate, replicate or improve on our results. Here, we present a follow-up project, focused on local authorities and cities with an EASI (Enable, Avoid, Shift and Improve) approach. We promote the establishment of a platform for the prevention of and response to marine pollution, to collect and exchange knowledge, while promoting the emergence of new behaviors from local stakeholders.

5.1. Introduction

The ISDAMP project took place from 2012–2015. Its aim was to help to prepare European countries to respond to marine pollution. The objective of this chapter is to present the dissemination of the results of the project and to suggest a possible follow-up project. We developed several decision aid systems using modeling and small-scale experiments.

The top management of partner institutions played a prominent role in the success of the project [DIN 16, pp. 57–73]. The operators' and territorial

Chapter written by Frédéric MUTTIN and Rose CAMPBELL.

authorities' knowledge of coastal pollution is also crucial. Their responses must follow a methodological guide, including both modeling and operational tools.

Project stakeholders included both external and internal bodies, and were considered in the project's scope and deliverables. Internal bodies included the project funders, research teams and top executives [ZWI 19, pp. 85–102]. Our main objective is the operational aspect of the project.

Our aim is to present the project's impact and continuation with a replicability purpose [AGU 20, pp. 45–66]. It consists of obtaining similar results on different data, using analogous methods. Future projects should apply a dual methodology over different coastal regions, by using both models and small-scale exercises.

We recall the ISDAMP project partnership:

– EIGSI is a French engineering school established in La Rochelle in 1991, and Casablanca, Morocco in 2005. EIGSI is a non-profit organization.

– IST is a Portuguese engineering school in Lisbon. IST is a public body and a component of the University of Lisbon.

– Action Modulers (AM) was the software developer of MOHID products (MOHID Studio). Action Modulers was a small for-profit company that was acquired by Bentley Systems in 2016.

– Falmouth Harbour Commissioners (FHC) is a statutory harbor authority in Cornwall, United Kingdom. FHC is a private UK-based company with a public-service delegation.

This chapter is organized as follows. First, we list the project deliverables. Second, we present the project results' audiences, from global to local institutions around the city of La Rochelle, France. Finally, we propose a concept for a follow-up project.

5.2. Project results

Table 5.1 defines the project deliverables list given at the beginning of the project, and indicates whether or not each deliverable has been produced. The last column gives the associated task. The lack of deliverable production could be the result of a cancelation, a modification or an addition of another deliverable. These reasons are not detailed.

The consortium agreement is the main result of the project management task. Several business models exist for knowledge interdependency [GIA 18, pp. 35–42]. We favor a sharing economy for environmental data and numerical software. Gathering partners' codes into a common source code was outside the objective, so we left out a collaborative economy.

The modeling task allowed improvements in IST's physical oceanographic modeling system and in EIGSI's floating barrier models. The IST system covers the Tagus Estuary and the Pertuis Charentais Sea in a wider Atlantic zone, ranging from Brittany to Morocco. The 1, 2 and 3D models of EIGSI are nested sequentially and use both ocean current maps and oil spill boom references as inputs.

The software task delivered two decision aid systems. The first is based on the coastal hydrodynamic MOHID-Studio tool and the second on the EIGSI BARRIER finite-element code. The Open-MI method permitted simultaneous 3D ocean hydrodynamics and 2D barrier structural analysis. Software documentation exists for both systems.

A mobile web application (Action Modulers) and a website giving real-time coastal oceanographic parameters on waves, winds and currents (IST) were available. During an emergency, an operator could launch the application on a mobile device. After the selection of a coastal area, an initial pollution location and a selection of boom anchors' positions, a screen displays the dynamics of the barrier and the oil spill in terms of sea current velocity and tidal water height.

We performed four exercises, instead of the three scheduled in the project scope. EIGSI purchased a modest stockpile of antipollution equipment, including both containment and sorbent boom sections. A GPS tracer and two small cameras track the boom position. A dynamometer and a recording device give the boom tension in real-time. A handheld anemometer and an infrared thermometer monitor *in situ* environmental parameters. Polyester mooring lines were used.

Training should share exercise videos and photos, as made by the Falmouth Harbour. There, a Tier 2 exercise is carried out every three years.

Large-scale exercises include subcontractors, local, regional and national authorities, observers, the media, maritime conferences and exhibitions. Mandatory information on exercises are human resources (company, number of operators, roles), equipment and costs.

	Yes	No	Task
Inception report (kick-off meeting in Brussels)	X		Management and reporting to the European Commission
Consortium Agreement	X		
Progress report no. 1 (technical and financial)	X		
Progress report no. 2 (technical and financial)	X		
Report on project evaluation		X	
Methodological guide (including transferability)		X	
Final technical implementation report	X		
Reports on hydrodynamics (3)	X		Modeling
Reports on barrier mechanics (3)		X	
Reports on model validation		X	
Functional requirements	X		Software
Water and barrier coupled modules	X		
Beta software versions, set of white-black box tests	X		
Software system documentation		X	
Reports on software testing		X	
Final software version	X		
Handbook with best practices		X	
List of sites for experiments	X		Exercises
Simulations report for each site retained		X	
Reports on experiments (3)	X		
Experiments final report (with cost analysis)		X	
Communication plan	X		Publicity
Website (isdamp.eu)	X		
Project brochures (2), newsletters (4), leaflets (2)		X	
First and second workshops	X		
Publications, articles, communications	X		
Training sessions (two per country)		X	
Layman's report (in English and French)		X	
Exploitation plan		X	

Table 5.1. *Available deliverables in terms of project task*

Task publicity fosters links between stakeholders. We have continuously updated social media (LinkedIn, Google+) and the ISDAMP website, with maintenance, up until 2020. The website publishes the communications made at the workshops (2014 and 2015 in La Rochelle) and the 2014 joint conference in Falmouth.

Our exploitation plan adapted a business model dedicated to research institutions [UNI 13]. The revenue stream addresses research grants. IST Lisbon had a continuation of its hydrodynamics topic through coordination with the HazRunoff project (2017–2019), funded by the European Commission DG-ECHO. Our business model key partners range from local to international.

In exercises and software tasks, Table 5.1 shows unachieved reports. This can be explained, in part, by an artificial increase of deliverables to satisfy the initial project evaluation [KAL 14]. The partners performed more than the scheduled small-scale exercises and decision aid systems. Nevertheless, numerous interpretations of expected tasks may have a negative effect on collective intelligence [BOU 16, pp. 45–98]. Knowledge management methods depend on the national culture of research teams [CHO 16]. These differences certainly impact the project and its follow-up. In the next section, we will develop the project dissemination to external stakeholders.

5.3. Project audiences

In this section, we enumerate the international, national, regional and local institutions interested by the project output.

5.3.1. *International institutions*

Table 5.2 details project activities within the worldwide maritime sector. The former AMSSA agency gave us expertise in maritime risk, training and experiments, together with Falmouth Harbour. The EMSA agency gave positive feedback for the Lisbon oil terminal experiment. The Atlantic action plan (February 3rd, 2015) and the CESER Nouvelle-Aquitaine, Réseau Transnational Atlantique (RTA) (November 24th, 2016) allowed knowledge sharing during two events on the transatlantic research perspective, organized in Paris. We communicated the project results during the IOSC 2013,

AMOP 2014 and INTERSPILL 2015 conferences. Action Modulers at the Working Group on Oil Spill Modeling Technologies WGOS first presented a clip showing MOHID hydrodynamics and BARRIER mechanics in 2015. The ARCOPOL competencies' platform [ARC 20] played a prominent role.

	Status	Organization	ISDAMP
Africa	International	African Maritime Safety and Security Agency (AMSSA)	Harbor exercises preparation
Europe	European Union agency	European Maritime Safety Agency (EMSA)	Lisbon harbor experiment
EU–US–Canada	Atlantic Ocean Research Alliance Galway 2013	Atlantic Action Plan	EUROMARITIME 2015, Paris RTA 2016, Paris
Worldwide	ARCOPOL-Platform project	Working Group on Oil Spill Modeling Technologies (WGOS)	ARCOPOL-Platform project at INTERSPILL 2015, Amsterdam
Worldwide	ARCOPOL-Platform project	Atlantic Technological Platform for oil and HNS spill preparedness, response and mitigation	Compile and promote knowledge and experience
Worldwide	Forum	IOSC conference, Savannah, USA, 2013	ISDAMP poster
Worldwide	Scientific forum	AMOP conference, Canmore, Canada, 2014	ISDAMP communication

Table 5.2. *International institutions relative to ISDAMP*

5.3.2. *National Contact Points*

In 2012, only Cyprus and France won a marine pollution project from the European Commission, ISDAMP and NEREIDs [ALV 15]. Generally, such projects include communication with all of the maritime contact points of European countries. Table 5.3 gives a list of such communications carried out during the project.

The numerous countries involved in the exercises, workshops, conferences and future projects are a positive aspect of the project. We shared our mobile application and the clip simulating an oil spill barrier, together with a polluted estuary. A tools assessment is still pending.

	Status	Organization	ISDAMP
Cyprus	National Contact Point	Centre of Operations – Civil Defense	NEREIDs DG-ECHO 2012 project
Finland	National Competent Authority	Finnish Environment Institute (SYKE)	Networking
France	National Competent Authority	Secrétariat Général de la Mer (SGMer)	Project Notification Agreement
Ireland	National Competent Authority	Irish Coast Guard, Department of Transport, Tourism and Sport	1st ISDAMP workshop 2014
Portugal	National Competent Authority	Directorate-General of Maritime Authority (SCPM/ DGAM)	Lisbon harbor exercise
Sweden	National Competent Authority	Swedish Coast Guard HQ – Marine Pollution response	1st ISDAMP workshop 2014
United Kingdom	National Competent Authority	Maritime and Coastguard Agency (MCA)	Presentation at the Falmouth 2014 ISDAMP-SPRES conference
Spain	National Contact Point for CECIS	Centro Nacional de Coordinación de Salvamento (CNCS) Sociedad de Salvamento y Seguridad Marítima (SASEMAR)	Exercises in Galicia and Andalucía ARCOPOL-Platform project

Table 5.3. *National Maritime Authorities relative to ISDAMP*

5.3.3. *Atlantic Arc regional authorities*

In the ARCOPOL-Platform project [ARC 20], the scaling-up of several project results began in two Spanish regions. The CETMAR institute, located in Vigo, led the project where IST, Action Modulers and EIGSI were involved.

We gave the structural analysis simulations of oil spill barriers installed during two exercises in the Galician Port of Caramiñal and the Andalusian Palmonés estuary. Table 5.4 presents three activities related to ISDAMP tools. We generalized them to protect regional parks.

Region	Activity	Specificity	ISDAMP
Galicia	Exercise at the Port of Caramiñal	Atlantic Ocean, Ría de Arousa, exercise on October 1st, 2014	BARRIER software for exercise simulations
Andalucía	Exercise at the Palmonés river estuary	Mediterranean Sea, North west of Algeciras Bay and Gibraltar Canyon, exercise on February 26th, 2015	BARRIER software for exercise simulations
Nouvelle-Aquitaine and Lisbon	Hydrodynamic modeling	MARETEC operational modeling tool	MOHID software for simulations of Pertuis Charentais and Tagus Estuary

Table 5.4. *Atlantic Arc regional organizations relative to ISDAMP through the ARCOPOL-Platform*

The European Commission DG-ECHO funded a continuation of ARCOPOL, the MARINER project (2016–2018). The CETMAR focused on responses to chemicals and Hazardous and Noxious Substances (HNS) spills.

5.3.4. *Local authorities in La Rochelle*

La Rochelle city belongs to the Charente-Maritime department. It has five islands, 56 harbors and 463 km of coastline. Our local government gave us a letter of intent, and shared resources for both the Rochefort and Chef de Baie harbor experiments. Table 5.5 details the local authorities' links to the ISDAMP project.

We give two examples of such links. The Pertuis Charentais hydrodynamics simulations are useful in the Pilot station's ship piloting simulations. The finite-element oil spill boom analysis is helpful for the development of POLMAR plans by Maritime Affairs, based on the finite-difference method FORBAR [FAU 95].

The commercial attractiveness of a harbor depends more and more on its approach to environmental issues. Pollution can be accidental or chronic. In the city of La Rochelle there are four harbors. During the ISDAMP workshops and exercises, we saw synergies between La Rochelle University, harbor environmental departments and EIGSI.

	Status	Organization	ISDAMP
Charente-Maritime Department	Local authority	Conseil Départemental Charente-Maritime	Letter of intent Rochefort Harbor and La Rochelle Chef de Baie experiments
La Rochelle Pallice Harbor Pilot Station	Pilot Station	Pilotage de La Rochelle-Charente	Rochefort Harbor experiment
Maritime Affairs of Charente-Maritime	POLMAR Plan	Direction Départementale des Territoires et de la Mer (DDTM)	Rochefort Harbor and La Rochelle Chef de Baie experiments
Rochefort Harbor	Commercial harbor	Harbor Master	Rochefort Harbor experiment, December 2014
Chef de Baie Harbor	Fishing harbor	Harbor Supervisor	La Rochelle Chef de Baie experiment, March 2015
Port de Plaisance des Minimes La Rochelle	La Rochelle marina	Environmental Department Manager	2nd ISDAMP workshop 2015
La Rochelle University	Research	LIttoral ENvironnement et Sociétés (LIENSs)	1st and 2nd ISDAMP workshops, 2014 and 2015 Intra-harbor water quality

Table 5.5. *Local organizations in La Rochelle relative to ISDAMP*

Before describing the follow-up project suggestions, the next section provides a brief discussion.

5.3.5. *Discussion*

The European maritime institutions' scope is more focused on policies and economic sectors than on individual projects. However, these institutions appreciate when project activities create synergies and share the best practices during experiments, workshops and conferences. The European projects that created such synergies with ISDAMP were NETMAR, ARCOPOL, SPRES, POSOW and NEREIDS.

The visions of European projects can be altered by insufficient knowledge of similar projects. Preliminary discussions between project participants and authorities could reduce the overlapping of projects. We need to do an ISDAMP exit strategy assessment with respect to the influence of similar projects.

While preparing an exercise, local stakeholders can share knowledge and discuss their roles in past incidents. Municipalities' policies use decision aid systems for flooding, transportation, energy and others. In the future, similar tools will be beneficial in response to pollution. For instance, basic data and simple mathematical models can be used to assist operators without heavy equipment, wearing protective gloves, to safely and efficiently handle short floating barriers and their moorings.

5.4. Project continuation

EIGSI's research expertise on local transport policies can be applied to coastal and marine pollution issues. The EIGSI research department suggested a follow-up of the project for local authorities, by using an Enable, Avoid, Shift, and Improve (EASI) approach for cities' integration of innovative pollution response methods. Several public transport policies used an EASI framework [GOR 17, pp. 127–149, DAL 19]. The final ISDAMP report contains another approach based on the Strength, Weakness, Opportunity and Threat analysis (SWOT), which is not detailed here [RIM 15].

Let us give the context of this follow-up:

– local authorities are in charge of marine and coastal pollution;

– apart from large disasters, they are left alone to cope with "day to day" pollution;

– they are not well prepared (lack of competences and resources);

– they are scattered along the coast and on several decision levels (region, department, city, harbor).

The new objectives will be to:

– combine local strengths and resources in accordance with the national level;

– increase the reactivity and agility of stakeholders;

– perform simplifications in modeling tools;

– reinforce the local ability to prevent, mitigate and respond to marine and coastal pollution.

Table 5.6 details the EASI summaries in both prevention and preparedness.

EASI	Enable	Avoid	Shift	Improve
Summary	Define resources, institutions' capacities and costs	Set up all elements to avoid local pollutions	Determine which stakeholders' behavior may be adapted	Improve existing tools to better respond to pollution
Efficiency	Resources	Prevention	Stakeholders	Preparedness

Table 5.6. *Marine pollution and EASI approach dedicated to local authorities*

We review the expected actions for each component.

Enable a local strategy that uses available resources to build valuable pollution policies. A collaborative platform would take inventory of the equipment for operators and local networks, through an information system. The platform could include data regarding environmental conditions and coastal infrastructures, for example, the shoreline and sea access morphologies. A city owning operational stockpiles could also investigate other risks, such as flood protection and waste disposal.

Avoiding local pollution could be encouraged by using different policies in transport and education about risks to the environment. The collaborative platform should store digital and physical images, maps and details of past pollution events. Showing simulations of response equipment and experiments under local oceanic and atmospheric conditions could foster local resilience. Avoiding significant amounts of pollution has a positive psychological impact.

Studying the ecological and societal perceptions of local representatives, bodies and companies could *shift* stakeholder behavior. Performing small-scale experiments allows participative science and the creation of innovative contingency plans. Maritime affairs, port authorities and the private sector usually perform Tier 1 exercises. Showing their results could transform practices and raise awareness amongst stakeholders.

Writing a manual of principles, which would be available on the platform, could *improve* stakeholder crisis communication. We could carry out a communication exercise on the toxicology data of the coastal environment. The environmental sensitivity and infrastructure vulnerability during past pollution events could be the subject of local consultations. The collaborative platform could illustrate simple pollution contingency strategies using makeshift equipment.

We stress that the documents on the platform must be standalone and easy to read, updated regularly, and available in both physical and digital formats. Tabletop exercises can use scenarios on small or large vessels powered by diesel or heavy fuel oil, respectively. During training at the Tier 1 level, we can test on-site access to electronic media and essential platform information.

We expect the impact to outperform ISDAMP. City networks' involvement favors the bottom-up approach between local and national levels. Such networks include the Conference of Atlantic Arc Cities (CAAC) and VIGIPOL, composed of cities in Brittany. Cities' local response plans should benefit from project outcomes. The collaboration of other projects performing small-scale exercises at the level of a city or harbor should deliver a valuable operational culture in risk management.

5.5. Conclusion

Throughout this chapter, we have involved both local to global institutions and organizations in our project. The proposed follow-up to the project focuses on the local level. Many questions will arise concerning communication about pollution, the collection of relevant information and citizen participation in ocean sciences, which are the subject of other European projects (xR4DRAMA, Blue-Cloud). A basic principle of communication during pollution events should concern all external stakeholders [ZIM 20, pp. 7–8]. With the help of the EASI framework, we believe we can impact the behavior of local stakeholders in the event of marine pollution.

5.6. Acknowledgments

We are grateful for the support of the European Union "Humanitarian Aid and Civil Protection", DG-ECHO, and the co-funding of the ISDAMP+ project by the Civil Protection Financial Instrument under grant no. 638516/2012/ECHO/A5/SUB.

5.7. References

[AFR 18] AFRICAN MARITIME SAFETY AND SECURITY, AMSSA, available at: http://www.amssa.net/, 2018.

[AGU 20] AGUINIS H., CASCIO W.F., RAMANI R.S., "Science's reproducibility and replicability crisis: International business is not immune", in EDEN L., NIELSEN B.B., VERBEKE A. (eds), *Research Methods in International Business*, Springer, Berlin, Heidelberg, 2020.

[ALV 15] ALVES T.M., KOKINOU E., ZODIATIS G. *et al.*, "Modelling of oil spills in confined maritime basins: The case for early response in the Eastern Mediterranean Sea", *Environmental Pollution*, vol. 206, pp. 390–399, 2015.

[ARC 20] ARCOPOL PLATFORM, Platform for improving maritime coastal pollution preparedness and response in Atlantic regions, CETMAR, Vigo, available at: http://arcopol.eu/, 2020.

[BOU 16] BOUVARD P., SUZANNE H., *Collective Intelligence Development in Business*, ISTE Ltd, London and John Wiley & Sons, New York, 2016.

[CHO 16] CHOUIKHA M.B., *Organizational Design for Knowledge Management*, ISTE Ltd, London and John Wiley & Sons, New York, 2016.

[DAL 19] DALKMANN H., "Air quality co-benefits of sustainable and low-carbon transport system: Future in Asia", *Intergovernmental Twelfth Regional Environmentally Sustainable Transport (EST) Forum in Asia*, Hanoi, Vietnam, October 28–31, 2019.

[DIN 16] DING R., *Key Project Management Based on Effective Project Thinking*, Springer, Berlin, Heidelberg, 2016.

[FAU 95] FAUVRE D., "Evaluation and limits of protective boom plans for high tidal range and strong current Areas", *Proceedings of the 1995 International Oil Spill Conference*, American Petroleum Institute, Washington, D.C. pp. 43–47, 1995.

[GIA 18] GIACOBBI V., "Economic models of knowledge sharing", in FABRE R., BENSOUSSAN A. (eds), *The Digital Factory for Knowledge: Production and Validation of Scientific Results*, ISTE Ltd, London and John Wiley & Sons, New York, 2018.

[GOR 17] GORHAM R., "Prospects for decarbonizing transport in Africa", in ISBELL P. and ÁLVAREZ PELEGRY, E. (eds), *Energy and Transportation in the Atlantic Basin*, Center for Transatlantic Relations, Washington, D.C., 2017.

[KAL 14] KALLENBACH S., JACQUET J., "La recherche partenariale, en France et ailleurs : des paysages en évolution constante", *Annales des Mines – Réalités industrielles*, vol. 1, pp. 67–88, 2014.

[RIM 15] RIM-RUKEH A., "Oil spill management in Nigeria: SWOT analysis of the joint investigation visit (JIV) process", *Journal of Environmental Protection*, vol. 6, no. 3, pp. 259–271, 2015.

[UNI 13] UNIVERSALIA, Business Models for Research Institutions, Universalia, Research paper, available at: http://documents.reflectlearn.org/Offline%20OA%20Tools/Business%20models%20paper_FINAL.pdf, August 2013.

[ZIM 20] ZIMMER J.V., Emergency response and crisis communication, ISCO Newsletter no. 732, International Spill Control Organization, London, 2020.

[ZWI 19] ZWIKAEL O., SMYRK J.R., *Project Management, A Benefit Realisation Approach*, Springer, Berlin, Heidelberg, 2019.

6

Environmental Quality of Coastal Areas in the Mediterranean Sea and Potential Risks to Human Health

6.1. Introduction

6.1.1. *Chemical pollution in coastal marine areas*

Human activities, whether at sea (e.g. fisheries and shipping), on land (e.g. agriculture and industries) or along the human-made coastline (e.g. wastewater treatment plants and massive tourism) may cause chronic and diffuse pollution in the marine environment [ISL 04, HOL 18] that represents 75% of the global marine pollution [PAR 20]. Seawater, sediments and organisms are contaminated by more than 100,000 substances and the list of new compounds is continuously increasing [MON 17]. In 2008, the European Commission approved the Marine Strategy Framework Directive [EUR 08a], a legislation aimed at achieving or maintaining "Good Environmental Status" in EU waters by 2020. In parallel to this directive, the European Parliament and the Council on environmental quality standards in the field of water policy defined Environmental Quality Standards (EQS) and published a list of 33 priority substances or groups of substances and eight other pollutants [EUR 08b], as well as their concentration limits, to

---

Chapter written by Marion PILLET, Michel MARENGO, Sylvie GOBERT, Pierre LEJEUNE, Michèle LEDUC, Lovina FULLGRABE, Stéphane LE FLOCH and Hélène THOMAS.

For a color version of all figures in this book, see www.iste.co.uk/muttin/marine.zip.

ensure good environmental status. The principal groups of chemicals targeted in this list are, among others:

– Polycyclic aromatic hydrocarbons (PAHs): of anthropic origin and mainly induced by the partial combustion of oil originated from urban activities (airplanes, ships, automobile, domestic heating) and industry (power generation plants or waste incinerators).

– Polychlorinated biphenyls (PCBs): derivatives of biphenyl compounds, these chemicals have been widely used in condensers and refrigerants, and can still be found in plasticizers, paints and adhesives [MEG 19]. Their main sources are linked to the collection and treatment of wastewater and that of the disposal and recovery of hazardous waste [INE 17].

– Polybromodiphenyls (PBDEs): used as flame retardants in, for example, houseware and textiles [KIM 17].

– Organochlorine pesticides (OCPs): these were extensively used worldwide from the 1940s to the 1970s for pest and insect control and are still being used in some countries for vector-borne disease control [VAN 17, CHO 11].

– Trace elements (TEs): they originate from natural processes such as rock weathering, wind-generated dust, river discharge and hydrothermal circulation at mid-ocean ridges [KRI 18], but also from anthropogenic processes such as mines and industries [CAS 15, STO 95], intensive agriculture, household waste and traffic [BUR 08, SEV 11]. TE pollution of the sea is less visible and direct than other types of marine pollution (e.g. macro-waste), but its effects on marine ecosystems and humans are intense and very extensive [ERK 07]. Some TEs are essential for life (e.g. selenium, copper, cobalt, molybdenum, manganese, zinc), with important biological functions that promote good health status in both marine organisms and people [VAN 15]. Essential elements could become toxic when present in quantities exceeding those required for accomplishing their biological functions [KRI 18, FRA 05]. However, other (non-essential) elements have no biological function (e.g. lead, cadmium, mercury, arsenic, tin, vanadium, aluminum), and become toxic when their intake is excessive or when ingested over a long time period. They can even be potentially carcinogenic to humans [OZK 17, LEC 19].

– Pharmaceuticals (e.g. analgesics, antihypertensives, antibiotics), personal care products (PCPs) (e.g. perfumes, insecticides, cosmetics) and

UV protection: directly used by people and end up in freshwater (rivers, lakes, groundwaters) and by extension in marine environments because they cannot be completely eliminated within wastewater treatment plants or directly as run-off from people's skin during recreational activities [PIC 14].

The contamination of the marine environment by these elements is problematic because they are non-biodegradable, toxic, have the potential to accumulate along the food chain and could act as endocrine disruptors [REN 13, PAR 20], causing adverse effects on the health of a living organism. However, the specific toxicity of most of these elements and their possible interactions (cocktail) remain unknown [SCH 09]. Thereby, it is crucial to develop a strong method to assess and monitor their diffusion in the environment, and also their potential effects on organisms in their pure form, or in combination with other substances [SVI 15]. For the above mentioned compounds and due to rapid dilution processes at sea, acute toxic effects are not expected. For the chronic toxicity or so-called "long-term effects", a biological response can be observed, even for a very low concentration, because chronic exposure typically induces a biological response of relatively slow progress and long duration in organisms.

6.1.2. *Bioindicators in marine coastal water*

Blandin [BLA 86] defined bioindicator species as "an organism or a group of organisms which, based on biochemical, cytological, physiological, ethological or ecological variables, allow the state of a given ecosystem to be assessed in a practical and dependable manner, while at the same time highlighting, as early as possible, the natural or induced alterations of the ecosystem". Evaluation can be measured at several levels, from the molecular and cellular levels to the populations up to whole biocenose levels [BON 15]. The use of bioindicators is particularly relevant from a biological point of view if the directly measured concentrations reflect the bioavailable fractions in the marine environment [WAN 16]. Marine biomonitoring consists of observing and assessing the current state and changes of marine ecosystems and provides information on the environmental impact of the pollutants.

In marine environments, a wide variety of species such as seagrasses [BON 18], fish [BEG 18], sponges [ORA 18], worms [BAR 18] and mollusks [PER 19, SOL 18, DIR 18, JI 19, PIN 15, VIÑ 18] are used as bioindicators. According to Azizi *et al.* [AZI 18a] and Rainbow and Phillips

[RAI 93], mussels would receive the greatest attention because they are widely distributed, very tolerant to environmental variations, easy to sample, of a reasonable size, sessile, present in high abundance, and sensitive to pollutants due to their filter ability, good bioaccumulators and ecological importance [LUN 15]. Classic examples of biomonitoring, using species of mussels and oysters as sentinel organisms are the Mussel Watch concept, which was first applied in 1975 [GOL 75, KRI 18, FAR 16] in order to conduct coastal chemical contaminant monitoring efforts in the United States, or the ROCCH network, initiated in 1974 [IFR 74], which consists of monitoring levels of chemical contamination of French coasts and their evolution over time. In principle, bioaccumulation is a function of bioavailable pollutant levels and is based on the absorption, the excretion and the accumulation of contaminants by mussel. But in practice, the concentrations in the mussel tissues depend on individual factors, such as the lifecycle, age and sexual maturity, and external environmental factors, such as food availability and salinity [AND 04, COS 89, PHI 76].

6.1.3. *Biomarkers used to monitor pollution*

The term biomarker refers to "all the biochemical, cellular, physiological or behavioral parameters which can be measured in the tissues or fluids of an organism or against the whole organism, to highlight the exposure to, or the effects, of one or more contaminants" [GAR 10].

At an individual level, changes in behavior can be considered as challenging to assess in bivalves. In the past, this integrative biomarker was not often used, due to the difficulty in assessing specific parameters, such as the shell growth [LEF 03], but new methods for measuring valve behavior responses [RED 17] or feeding efficiencies via measurements of the ciliary motility [CAP 15] were demonstrated as sensitive parameters in mussels. Stress on Stress (SoS), calculated on the basis of the survival time when organisms are exposed to air, is also a whole organism response, and can be used as a physiological index of health [MAR 17].

At a subcellular level, biomarkers can be divided into two groups: (1) biomarkers of exposure, which indicate the presence of a contaminant inside the organism, due to an interaction of this contaminant with biological molecules, and (2) biomarkers of effect, which indicate the toxic or sublethal effect of the contaminant on the organism [LUN 17]. Biomarkers of effect are chosen according to the targeted physiological functions (e.g. oxidative stress, immunity, etc.) and the impact of pollution on them.

Oxidative stress is commonly studied in ecotoxicology. It is defined as an imbalance between oxidant production and antioxidant capacity. In normal conditions, the antioxidant defenses are efficient enough to neutralize the natural production of reactive oxygen species (ROS) as side products of aerobic metabolism [HAL 99, LIV 01], and the health of the organisms is therefore maintained. In adverse conditions, when organisms are exposed to stress conditions such as pollution, the ROS levels increase and this crucial balance might be disrupted, leading to osmoregulatory dysfunctions, tissue damage via the oxidation of proteins and lipids, alteration of gene expression [LUS 11, RAD 88] or apoptosis [PEL 09]. Antioxidant defenses, characterized by enzymes and antioxidant compounds, are present to prevent ROS from causing adverse effects [GAG 14]. Three enzymes deal directly with ROS: superoxide dismutase (SOD) participates early in the antioxidative process by scavenging the superoxide radicals (O_2–•) to molecular oxygen (O_2) and hydrogen peroxide (H_2O_2) [MCC 69]. Catalase (CAT) and glutathione peroxidase (GPx) convert H_2O_2 and lipid hydro-peroxides in H_2O and O_2 [WEY 10]. Other enzymes contribute to the maintenance of the level of cellular glutathione, another antioxidant: glutathione reductase (GR) catalyzes the reduction of glutathione disulfide (GSSG) to maintain the GSH/GSSG ratio and glutathione S-transferases (GST) is implicated in the initiation of the detoxification of xenobiotics and toxin biotransformation [HAB 74]. Through the reduction of hydrogen peroxides and other hydroperoxides, the glutathione system provides protection against oxidative damages and the accumulation of free-radical products [TAP 82].

Perturbation in lipid metabolism is also known as a common and general stress response [BER 05, DAI 11]. First, cellular membranes can be damaged due to lipid peroxidation [ROC 14]. These damages are often observed via malondialdehyde (MDA) analysis. Second, fatty acid levels [LOA 20], an energetic metabolism and energy reserves [ADA 15, FAL 19] can be disturbed by pollutant exposure.

Immune systems can be impacted in organisms facing chronic and diffuse pollution. Phenol oxidases (POs) are a family of enzymes including tyrosinases, catecholases and laccases that are involved in the immune defense mechanisms in various invertebrates, like self/non-self-recognition, phagocytosis and nodule and capsule formations [ASO 97, SÖD 98, LUN 11].

The nervous system can be affected by environmental parameters and acetylcholine esterase (AChE), involved in the transmission of nerve impulses, and is used as an indicator of neurotoxicity [MAT 05].

Confronted with pollution, organisms display detoxification abilities and metallothionein (MT), a family of metal-binding proteins which play a key role in detoxification process, especially in the case of metal pollution. They are already used as biomarkers in organisms used for monitoring programs [AMI 06].

All of these biomarkers can present natural, environmental and biological variations [MON 07, KAA 99, ROB 03, VID 02].

6.2. Situation in the Mediterranean Sea

6.2.1. *Characteristics of the Mediterranean system*

The Mediterranean Sea is an enclosed basin with an average depth of 1,500 m, with short ventilation and residence time. The Mediterranean Sea has relatively warm and salty waters, due to the Mediterranean climate (defined by warm-dry summers and mild-humid winters) and its enclosed nature. It is divided into different sub-basins (the Alboran Sea, Balearic Sea, Liguro–Provencal and Catalan sub-basins, Algerian Sea, Ligurian Sea, Tyrrhenian Sea, Adriatic Sea, Ionian Sea, Levantine Sea and Aegean Sea), each of them characterized by different geomorphological and hydrological features. The Mediterranean displays oligotrophic characteristics, namely a low concentration of nutrients in the 100 m deep photic layer [BET 80]. Nutrient concentrations depend on the exchanges through the Strait of Gibraltar and the Bosporus, on the atmospheric depositions and on the river discharges [DUR 11]. The Western part of the Mediterranean Sea, connected to the Atlantic Ocean through the Strait of Gibraltar, is more productive than the Eastern part, due to the influence of the Atlantic Ocean.

In the Mediterranean Sea, the primary production is locally enhanced by freshwater inputs, but globally limited by the oligotrophic nature and strong solar radiation [DUR 11, CAH 72]. Generally, the biomass of phytoplankton, dominated by diatoms and phytoflagellates, starts to increase in February–March, until it reaches a peak in the upper water layers in spring [RIB 04, FUL 20]. As the Western basin of the Mediterranean Sea is more productive than the Eastern basin, due to the influence of the Atlantic Ocean, a west to east gradient in primary producers (phytoplankton) [BOS 04] is also reflected in the abundance of their primary consumers (zooplankton). The primary producers are predated first by primary and then by secondary pelagic consumers. The west-to-east gradient in productivity is reflected in zooplankton abundances [BER 12]. However, a coastal–open ocean gradient might dominate over such a basin scale gradient, leading to coastal site-specific planktonic characteristics [MAZ 07].

The epipelagic zooplankton of the Mediterranean Sea is characterized by diversified copepod assemblages that originated from both temperate and subtropical regions [RAZ 91, MAZ 03]. Multi-year studies conducted in the different basins demonstrated that these patterns in community change were influenced at seasonal and decadal scales, mainly in response to climate forcing [MAZ 07, FUL 20, DUR 11, RIB 04, FER 08]. The observed increase in water temperature and salinity was linked to species abundance changes since the 1950s [KOU 98, MAZ 07, VIL 20]. However, important anthropogenic forcing (increasing demographic and economic developments) in the Mediterranean basin also altered planktonic assemblages and abundances [DUA 99, KOU 98, BER 12, MAR 05]. Moreover, the oligotrophic nature of the Mediterranean Sea makes it more sensitive to anthropogenic forcing and especially to bioaccumulation processes because contaminants are less diluted by organic carbon [DUR 11].

The Mediterranean Sea, with densely populated coasts, is under numerous anthropogenic pressures. The pollutants, coming from the continental activities, are transported through rivers or by air and accumulate along coastal areas. The Mediterranean is one of the richest regions of Europe in terms of the diversity of marine species, with a high rate of endemism [BIA 00]. Pressures suffered by the Mediterranean make it a vulnerable ecological unit [TUR 99]. This sea is of too small dimensions to ecologically self-counterbalance. The point of saturation of the pollutants discharged in the Mediterranean, whose area is a 35th of that of the Atlantic, will be more quickly achieved than in the oceans. In addition, the almost total absence of tide does not allow the dilution of pollutants and prevents the natural phenomena of depuration, as encountered in larger bodies of water (i.e. in oceans). The Mediterranean also shows a deficiency in the movement of its deep water masses, and of its surface currents which "turn in circles" in this almost closed basin. Also, the Mediterranean's response to environmental disturbances is more rapid than in the larger oceans

[AUG 10]. This site could be used to predict the scenarios that we could live in the future, in all of the oceans of the world [BET 99, RIC 12].

6.2.2. *Contamination at the lower trophic levels*

The contamination at the lower trophic level of the food chain is difficult to evaluate using planktonic organisms. Changes in planktonic communities and planktonic species abundances have occasionally been used as an indicator to investigate the impact of pollution in marine communities [SIO 91, SER 16, TET 08, URI 04]. But few studies have used the direct measurement of contamination. For example, the TE bioaccumulation was evaluated in phyto- and zooplanktonic communities [BAT 17, ROS 08]. Despite the abilities of phyto- [ROS 08] and zooplankton [BAT 17] to accumulate metals, the sampling effort, small size, diversity and short lifetime of these organisms, challenge their use for bioaccumulation study.

To have an indication of the contamination at low trophic levels, seagrass or macroalgae can be privileged. The longer lifespan of these organisms can reflect integrated changes over a long period of time [DUA 99]. In the Mediterranean system, several species have proven their ability to concentrate high levels of pollutants, especially TE [BON 18, PER 00]. In the western basin, the dominant seagrass, *Posidonia oceanica* (L.) Delile is the most studied organism [GOB 09, CAT 93, PER 98], due to the vast area it occupies. *Posidonia oceanica* has been successfully used in an index to assess the ecological status of seawater along French Mediterranean coasts, showing that the quality of seawater was negatively correlated with the anthropization index [GOB 09]. The levels of the seven widely studied TE (Cr, Fe, Ni, Cu, Zn, Cd and Pb) seem to decrease or stabilize over time, probably due to their reduced anthropogenic inflows [RIC 12].

Globally, seagrass reflects the amount of pollutant in the surrounding seawater and sediments, but some issues persist. First, only a few different species were investigated and second, most studies only focus their research on TE contamination because seagrass does not bioaccumulate organic pollutants well [BON 18].

6.3. Mussel biomonitoring in the Mediterranean Sea

6.3.1. *Biomonitoring of mussels*

To obtain complete information about a system's contamination, the use of filter organisms, and especially mussels, are generally privileged. They reflect the contamination at a low trophic level, but are also consumed by larger organisms (including humans) and give direct indication of potential contamination at a higher trophic level or the potential human risk. Because of this phytosanitary aspect, the health status of the coastal Mediterranean should be monitored in mussels [RIC 14]. Marine mussels, as filtering organisms, are useful for biomonitoring and assessing the current and historical state of marine contamination [e.g. JOH 06, MIL 16, ULU 19, AND 04, BEN 11, AZI 18b]. Mussels can uptake contaminants via different mechanisms: passive diffusion over external surface and active uptake by transmembrane ion pump transport, or by endocytosis of contaminated particles. Accumulated concentrations are dependent on the physicochemical properties of the contaminant, its bioavailability in the water and physiological factors related to the organism [BEY 17].

The Mediterranean mussel (*Mytilus galloprovincialis* L., 1819) is a key component of the biogeochemical cycle in the marine system and is one of the most important Mollusca of the Mediterranean Sea and the Black Sea [YUA 20]. They have an important role in the recycling of nutrients and particles, using suspended materials from the water column [BAT 18]. The Mediterranean mussel is a shellfish species that is widely distributed in the coastal waters of the eastern Atlantic-Mediterranean waters [YIG 18]. Their usefulness as sentinel organisms in TE biomonitoring studies is widely recognized [RIC 14]. Due to their tolerance to diverse environmental conditions, their sedentary lifestyle and their ability to accumulate various TEs, mussels are considered reliable bioindicators of marine ecosystem pollution [BOG 14].

There are two ways of biomonitoring with mussels, first, passive biomonitoring using wild mussels, and second, active biomonitoring by transplanting wild mussels from a reference site or cultivated mussels directly from hatcheries in caging. The advantage of the first is the real representation of environmental perturbations during time exposure, but its main problem is the scarcity of natural mussel stock in several coastal areas. The advantages of the active biomonitoring are the total control of the depth and location of caging. Active biomonitoring also allows the reduction of biological individual variability (age, size) by using control organisms from hatcheries. But, using this method at a large scale increases the influence of external factors, such as shell growth, which can be really affected by hydrodynamism [AND 04, REG 93].

6.3.2. *Trace element contamination of mussels in the Mediterranean Sea*

The accumulation of trace elements depends on the exposure routes (diet or surrounding water/sediment) and on the metal speciation, related to the physicochemical parameters of the water [BEY 17, GOL 80, AZI 18b]. However, several studies demonstrate that the bioaccumulation pattern in mussel tissues are usually Fe > Zn > Mn > Cu > Ni > Cr > Pb > Cd (Table 6.1) [ULU 19, AZI 18b], indicating a higher tendency for essential metals [RAI 02]. For mussels, the temporal evolution of TE accumulation usually follows an asymptotic curve, until a steady-state with the surrounding water is reached within 2–3 weeks [RIC 14]. The levels of TE accumulation in wild mussels and in caged mussels are identical [AND 04, CAR 19, ABD 19].

The variability in *M. galloprovincialis* TE accumulation usually follows a seasonal cycle related to many factors, such as variations in seawater temperature, food availability and ecophysiological functions (e.g. growth, reproduction or constitution of reserve substances) [RIC 16a, ULU 19, AZI 18b, BEN 15]. The highest TE concentrations are generally observed in winter [ULU 19, AZI 18b, BEN 15], even if some local differences can be observed, for example in the Adriatic Sea where the highest Hg concentrations of mussels are observed in September [BAJ 19]. This increase of TE could be related to the higher bioaccumulation rate of the organism during the pre-reproductive period and the developing gonadal tissues [ULU 19, RIC 14]. It is important to note that this increase can differ between males and females [RIC 16b]. However, the impact of spawning on the trace metal level in mussels is still up for debate. While some studies hypothesize that the emission of gametes provokes steep increases of TE due to weight loss [RIC 14], others suggest that gonads may constitute an effective trap for the incorporation of TE and so the emission of gametes would be accompanied by a fall in the TE content [AZI 18b].

Tissue	Zn	Ag	As	Al	Fe	Cd	Pb	Mn	Mo	Hg	Cu	Ni	Cr	Co	V	Sb	Sn	Se	Pt	Location	Reference
Soft tissue	120-200	-	10-69	-	-	0.55-5.8	0.5-5.3	-	-	0.04-0.34	3.1-9.3	1-10.5	0.5-7.4	-	-	-	-	-	-	French Mediterranean coast	a (*caging*)
	-	-	-	-	-	-	-	-	-	-	-	-	-	-	-	-	-	-	0.0001-0.0007		b (*in situ*)
	66.2-79.7	0.01-0.16	26.9-32.7	150-323	146-255	0.35-0.39	0.26-0.4	6.9-12.9	12.3-16.3	-	3.6-5.5	1.16-1.81	0.5-0.8	0.60-0.68	4.1-5.8	0.011-0.015	0.02-0.04	2.2-3.2	-		c (*caging*)
	128-394	-	-	-	-	0.55-1.01	-	-	-	-	3.1-202.3	-	-	-	-	-	-	-	-		d (*caging*)
	117-121	0.01-0.02	12.8-14.3	25-55	74-101	0.71-0.8	0.84-0.99	2-2.5	4.9-5.5	-	2.8-3.3	0.85-1.02	0.2-0.32	0.43-0.52	1.3-1.5	0.01-0.014	0.01-0.02	2.9-3.4	-	Corsica coast	e (*caging*)
	-	-	-	-	-	1.13-1.82	1.07-1.43	-	-	0.09-0.12	-	0.1-3.67	0.4-3	0.06-1.43	-	-	-	-	-	North-western Mediterranean	f (*caging*)
	95-186	-	-	-	-	0.1-0.1	0.64-4.28	-	-	0.02-2.36	6.4-32.3	-	0.03-1.6	-	-	-	-	-	-	Adriatic Sea	g (*in situ*)
	94-174	-	-	-	-	0.45-1.03	0.5-1.5	-	-	0.05-0.12	2.7-10.1	0.6-2.9	0.3-2.4	-	1.2-10.2	-	-	-	-		h (*caging*)
	160-164	-	-	-	340-673	0.64-0.85	< LOQ	-	-	-	5.2-6.2	2.03-3.82	1.9-3.3	< LOQ	-	-	-	-	-	Northern Moroccan coast	i (*caging*)
	114-255	-	-	211-1032	241-553	1.36-2.01	1.94-2.29	0.62-3.55	-	0.21-1.53	1.9-3.7	2.86-3.45	4.1-6.2	-	-	-	-	-	-	Sicilian coast	j (*caging*)
	71.3-184	-	-	-	56.1-349	0.22-0.59	0.41-1.54	4.37-12.7	-	0.07-0.17	2.1-6.3	-	1.1-6.63	-	-	-	-	-	-	Eastern Aegean Sea	k (*in situ*)
Dig. gland	41.1-142	-	-	-	-	0.6-1.8	< LOQ-0.4	5.2-46.8	-	< LOQ	2.4-16.4	2.9-7.9	0.5-8.6	-	-	-	-	-	-	Southern Greece	l (*caging*)

Table 6.1. *Example of the bioaccumulation of trace elements (min-max of mean values; expressed in µg/g of dry weight) in Mytilus galloprovincialis using data from recent studies: (a) [AND 04]; (b) [ABD 19]; (c) [RIC 16b]; (d) [ROM 03]; (e) [RIC 14]; (f) [LAF 07]; (g) [TSA 16]; (h) [BAJ 19]; (i) [AZI 18b]; (j) [CAR 19]; (k) [ULU 19] and (l) [PYT 08]. LOQ = Limit of Quantification*

Finally, it should not be forgotten that the higher TE concentration reported in spring could also be linked to an increase in the seasonal change in anthropogenic activity during the touristic period, due to there being more marine shipping and recreational activities [RIC 14, AND 04].

Spatial variations of TE accumulation patterns are also observed. Globally, in the Mediterranean Sea, TE levels do not exceed the established quality standards recommended by various organizations [AZI 18b, RIC 14] and they follow an increasing gradient of concentration from north to south and from west to east (Table 6.1) [ABD 19]. However, taking the TE individually, there are spatial differences. For example, Cu concentration reported in mussels from the Adriatic Sea were lower in the Adriatic Sea and the Libyan coast in comparison to the Turkish coast [BAJ 19]; the Western Mediterranean Sea presents lower levels of Ni compared to the Adriatic Sea [BEN 11]; the lowest Zn concentrations were reported in the Adriatic Sea [BAJ 19]; and the highest Cd concentrations were observed in mussels from the Libyan coast, compared to the Western basin or the Adriatic Sea [BAJ 19, BEN 11]. At a smaller scale, lagoons linked to the intensity of human activity in the associated watershed or location close to major urban and industrial centers are more contaminated compared to open-sea regions [AND 04, ULU 19, MIL 16, ABD 19]. Moreover, some local sites can present high contamination of specific TEs due to a local source of Cd contamination. For example, in the north-west of Corsica, high levels of Ni and Cr reveal the presence of old asbestos mines [AND 04].

6.3.3. *Organic contamination of mussels in the Mediterranean Sea*

The uptake and detoxification of hydrophobic organic contaminants (PCBs, PAHs, etc.) happens mostly by passive diffusion, the sources of contamination being chemicals dissolved in seawater and food particles [GOL 80]. The intake from food is more important than from water because the concentrations of lipophilic compounds in the dissolved phase are very low. Therefore, suspended organic matter, phytoplankton that bioconcentrates lipophilic compounds directly from the water [BER 00], and zooplankton may be sources of lipophilic contaminants for mussels. The accumulation of contaminants is dependent of biological factors (age, lipid content, reproduction, etc.) and environmental factors (properties of the contaminant itself, seasonal changes, organic matter, primary production, precipitation, etc.) and of the metabolic detoxification abilities of the organism [MIL 16,

KWO 09]. Studies showed that even if mussels respond rapidly to changes in environmental conditions [MIL 16], it can take two to three months to attain a steady state between the organism and the environment [GAL 11].

In previous studies all over the Mediterranean Sea (Table 6.2), authors demonstrated that the seasonal and spatial variations of organochlorine pollutants were the result of prevailing environmental, rather than biological, parameters [MIL 16, ULU 19].

Polycyclic aromatic hydrocarbons concentrations detected in marine mussels are higher in coastal areas, which are usually nearer to the main cities and ports and are affected by maritime traffic [LEÓ 13, BAJ 19]. In the Mediterranean system, the higher values were recorded along the Iberian [LEÓ 13] and French coasts [AND 04] and in the Aegean Sea [TSA 16], and the lower concentrations in mussels were recorded in the Adriatic Sea [BAJ 19]. These concentrations are also lower than those commonly detected in mussels from the Atlantic Ocean and North Sea, probably due to the higher maritime traffic in those areas and the lower temperatures that reduce PAHs volatilization [LEÓ 13]. The PAHs levels also display a temporal variation, with observed values increasing with the development of port activity and population. Indeed, the PAHs composition in the Mediterranean Sea reveals that phenanthrene is the most abundant congener. The PAHs have a mixed origin, they could reflect a significant contribution of petrogenic processes and could be the by-products of the detoxification process by organisms [BAJ 19, LEÓ 13].

From previous studies, higher levels of PCBs were recorded on the French coast sites [AND 04] and in the Adriatic Sea [MIL 16, BAJ 19]. PCBs originated mainly from commercial mixtures, such as Aroclor 1254 or 1260, and are commonly used in European countries. The accumulation of higher PCBs was recorded in areas highly influenced by wastewater discharge and mainland run-off, in marinas and in bays where the low hydrodynamism and slow exchange of water reduce the dilution of the pollutants [BAJ 19, GAL 11]. The accumulation pattern in the mussels also reflects the composition of PCB congeners present in the ambient water [MIL 16, BAJ 19]. PCB congeners with more than five chlorine and para-chlorine atoms are eliminated much slower than less chlorinated congeners. The levels and distribution of PCB congeners in mussels are highly marked by seasonal variation, which could be due to the physiological variability of organisms [MIL 16, BAJ 19].

Tissue	Σ PAHs	Σ PCB	PBDEs	POPs	DDE	Location	Reference
M. galloprovincialis soft tissue	0.4-17.2	5-10.8	-	< LOQ-0.011	< LOQ-0.075	French Mediterranean coast	a (*caging*)
	0.11-4.72	0.008-0.019	-	-	-	Northern Adriatic Sea	b (*caging*)
	0.03-0.51	0.008-0.013	-	-	-	Adriatic Sea	c (*in situ*)
	0.04-0.35	0.010-0.39	-	-	-	Aegean Sea	c (*in situ*)
	0.5-6.1	4-115	-	0.02-0.04	< LOQ-0.012	Coastal waters of the Adriatic Sea	d (*caging*)
	-	-	-	-	< LOQ-0.071	Eastern Aegean Sea	e (*in situ*)
	5-389.5	-	-	-	-	Iberian Mediterranean coast	f (*in situ*)
	-	0.81-23.2	-	0.014-0.04	-	Eastern Adriatic Sea	g (*caging* and *in situ*)
	0.037-0.05	-	-	-	-	North western Mediterranean basin	h (*caging*)
M. galloprovincialis and *M. edulis* soft tissue	-	-	< LOQ-0.00008	-	-	French Mediterranean coast	i (*in situ*)
M. galloprovincialis hepatopancreas	-	-	-	-	< LOQ-0.02	Eastern Aegean Sea	e (*in situ*)

Table 6.2. *Examples of organic pollutants bioaccumulation (min-max of mean values; expressed in µg/g of dry weight) in mussels using data from recent studies: (a) [AND 04]; (b) [MOS 16]; (c) [TSA 16]; (d) [BAJ 19]; (e) [ULU 19]; (f) [LEÓ 13]; (g) [MIL 16]; (h) [GAL 11] and (i) [JOH 06]. POPs = Persistant Organic Pollutants; DDE = DichloroDiphenyldichloroEthylene (by-product of dichlorodiphenyltrichloroethane DDT); PBDEs = PolyBrominated Biphenyl Ethers; PAHs = Polycyclic Aromatic Hydrocarbons; LOQ = Limit of Quantification*

Few studies focus on the PBDE accumulation, but the levels recorded in the Mediterranean Sea are much lower than the ones on the Atlantic or in the English Channel [JOH 06]. As for the other pollutants, the PBDE concentrations reflect the influence of urban and industrial sources in coastal areas.

Persistent organic pollutants (POPs), such as "drins" compounds or dichlorodiphenyltrichloroethane (DDT), are found in the marine areas near agricultural and industrial activity sites, due to their past use as a pesticide and insecticide. The levels of "drins" compounds and DDT in the Mediterranean Sea mussels reflect how popular they were in the 1960s, as well as the slow decrease that took place due to them being banned in many countries [AND 04, ULU 19, BAJ 19]. Spatially, the DDT concentrations decrease from the north-western Mediterranean to the southwestern Mediterranean [BAJ 19], but these levels vary according to the studied sites with, for example, a high concentration in the Mar Menor Lagoon in Spain [LEÓ 13] and levels below the limit of quantification in the central-south Italian coasts [BAJ 19]. Dichlorodiphenyldichloroethylene (DDE), as a degradation product of DDT, can stay in the environment for years without degradation [MIL 16] and is sometimes found at high levels, especially in the coastal areas of the French Mediterranean Sea [AND 04] and in the Aegean Sea [ULU 19], where agricultural run-off, urban discharges and harbor activities are prominent [BAJ 19]. Contrarily, the DDE contamination in the central-south Italian coasts was moderate [BAJ 19].

Globally, the recorded organic pollutant concentrations in the Mediterranean Sea were below the prescribed limits for human consumption and were not considered as a risk to human health [MIL 16, ULU 19]. The accumulation was higher in the coastal area impacted by anthropogenic pollution [BAJ 19, GAL 11] rather than in the open-sea regions where there is a dilution effect and a less particulate matter from watersheds [AND 04]. However, long temporal series are required to accurately determine the environmental changes [LEÓ 13, BAJ 19].

6.3.4. *Physiological functions affected by pollution*

Facing pollution, several physiological functions of mussels can be impacted. So, it is important to also measure the biomarkers of effects to evaluate the health of organisms.

The antioxidant defenses, estimated as good biomarkers [NAS 02, FIT 93], are usually the first ones to be analyzed. Mollusks generally present quite high activity of antioxidant enzymes [BEN 15, PAM 05, VAL 06]. These enzymes showed quick responses to stress in general. For example, SOD activity increased in response to metals and many other pollutants [FER 10] and an increase of GST activity was observed in polluted areas [BEN 15]. But sometimes, the responses are not pollutant specific [BEN 15, DAM 04, ROM 03] and can show opposite responses according to pollutant concentration or duration exposure. While an inhibition of CAT activity has been evaluated as a temporary response to acute pollution [OZK 17, REG 95, DUA 11], positive correlations have been found between its activity and the concentration of TE [DUA 11] or mixture of pollutants such as tributyltin, PAHs and PCBs in harbors [BEN 15, ROM 03, VEN 02, VIA 91]. Moreover, in long exposure conditions to high levels of pollutants, some studies showed that the antioxidant defenses can be overrun [DUA 11, LIV 91, LOW 00, REG 95, STO 95]. However, in general, in environmental exposure, mussels are able to adapt and the activities of antioxidant defenses go back to normal [LIO 03].

The lipid and fatty acid composition of mussels can be perturbed and a lot of changes are observed in response to chemical pollution. Lipid peroxidation increased [OZK 17, ROM 97, SRI 98], changes in the structure and fluidity of the cell lipid bilayer [FOK 13] caused lysosomal storage of neutral lipids [DAI 11], decreased the lysosomal membrane stability and widened the lysosomes when there is an exposition of marine organisms to many pollutants [MOO 04].

Other physiological functions, despite being less studied, can be impacted by exposure to contaminants. The immune defense mechanisms in various invertebrates, via variation in phenoloxidases activity, can be disturbed, leading to cascade reactions, and sometimes inducing the destruction of microorganisms [LUN 11]. Some pollutants can cause neurotoxic damage, as was demonstrated by the inhibition of AChE activity in the presence of different contaminants like organophosphate and carbamate pesticides [LIO 13], TE and hydrocarbons [CAP 15] and mixtures of pollutants [DAG 14, MAT 05].

6.4. Potential human risk assessment

6.4.1. *Seafood, potential risks and benefits for human health*

Fisheries and aquaculture represent significant resources for hundreds of millions of people around the world, whether as a source of food, income or livelihoods [FAO 18]. The average global seafood consumption per year has now reached a record of 20 kg per person, reflecting the high potential of this resource to participate in global food security [DUP 20]. Thus, in recent years, there has been an increase in global seafood consumption, mainly due to human health benefits [LEI 13]. Marine products are a good source of proteins and lipids of high nutritional value, with long-chain polyunsaturated fatty acids (Omega-3), liposoluble vitamins, essential elements that can make a positive contribution to the prevention of cardiovascular diseases [DIB 15]. In spite of these advantages, the accumulation of toxic TE in marine species can counteract their beneficial effects and may pose a potential risk to human health [MED 12]. There is a growing interest in food safety with regard to the potential accumulation of these contaminants and their consequences [LEI 13].

The consumption of TEs can cause adverse effects, for example, renal dysfunction, lung disease, liver failure, anemia, reproductive deficiencies, premature births and chronic damage to the central and peripheral nervous system [DAD 16, SQU 16]. These elements, due to their persistence, bioaccumulation and possible biomagnification in the food web, represent a serious problem for the environment and human health [DEM 06]. Trace elements can be bioaccumulated by marine organisms through a variety of pathways, including through respiration, adsorption and ingestion [TÜR 08]. Several factors can influence the bioaccumulation of TEs, such as environmental conditions (temperature, salinity, pH, etc.), biological variations (species, sex, size and age), food source, habitat, type/level and time of exposure to pollution [MAR 18]. These elements can be biomagnified via the food chain and end up assimilated by people who consume marine food, which poses a risk to their health [BAE 05]. Thus, marine species represent a powerful model for risk–benefit assessment [DIB 15].

6.4.2. *Potential risks associated with mussel consumption*

The production of mussels, including through both aquaculture and wild harvesting, is increasing around the Mediterranean. The main producers are Spain, France, Italy, Algeria and Morocco [ÖZD 10]. In order to protect the health of seafood consumers, many directives and international organizations have defined limits on permissible concentrations of toxic pollutants in seafood [JEC 11, EUR 06]. To assess the potential risk to human health, several indices have been developed, for example the Estimated Weekly Intake (EWI), Total Hazard Index (THI) and Maximum Allowable Daily Consumption Limits (MADCL).

Analyzed trace elements	Human health indices	Potential health risk to consumers	Location	Region	Reference
Cd, Cr, Cu, Fe, Hg, Ni, Pb, Zn	EWI, THQ	No risks	Greece	Mediterranean Sea	a
As, Ba, Cd, Co, Cr, Cu Fe, Hg, I, Mn, Ni, Sb, Sn, Sr, V, Zn, Zr	THQ	No risks	Montenegro	Mediterranean Sea	b
Al Cd Cr Co Cu Fe Hg Mn Ni Pb Sr Zn	THI, THQ	No risks	Montenegro	Mediterranean Sea	c
As, Ba, Br, Ca, Cd, Ce, Cl, Co, Cr, Cs, Cu, Fe, I, K, Mn, Ni, P, Pb, Rb, S, Sb, Si, Sn, Sr, Th, Ti, V, Zn, Zr	EPI, THI, THQ	No risks	Montenegro	Mediterranean Sea	d
Cu, Fe, Mn, Zn	EDC, IRR, MADCL, MACR, THQ, THI	No risks	Turkey	Mediterranean Sea	e
Cd, Hg, Pb	EDI, THQ, THI	No risks	Bulgaria	Black Sea	f
Cd, Cu, Fe, Hg, Mn, Pb, Zn	EDI, EWI, THI	No risks	Turkey	Black Sea	g
Cd, Cr, Cu, Mn, Ni, Pb, Zn	THQ, EDI	Risks	Turkey	Black Sea	h
As, Cd, Cr, Cu, Fe, Hg, Mn, Ni, Pb, Zn	THQ, HI, TR	No risks	Bulgaria	Black Sea	i

EDC = Estimated Daily Consumption; EDI = Estimatd Daily Intakes; EPI = Element Pollution Index; EWI = Estimated Weekly Intake; HI = Hazard Index; IRR = Individual Risk Ratio; MACR = Maximum Allowable Consumption Rate; MADCL = Maximum Allowable Daily Consumption Limits; THI = Total Hazard Index; THQ = Target Hazard Quotient; TR = Target Risks

Table 6.3. *Comparison of human health risk assessment for consumers, based on trace elements, in Mytilus galloprovincialis from the Mediterranean Sea and Black Sea coasts, analyzed with literature data: (a) [KAL 12]; (b) [JOV 15]; (c) [PER 18]; (d) [TAN 16]; (e) [YIG 18]; (f) [ZHE 18]; (g) [BAT 18]; (h) [GED 17] and (i) [PEY 19]*

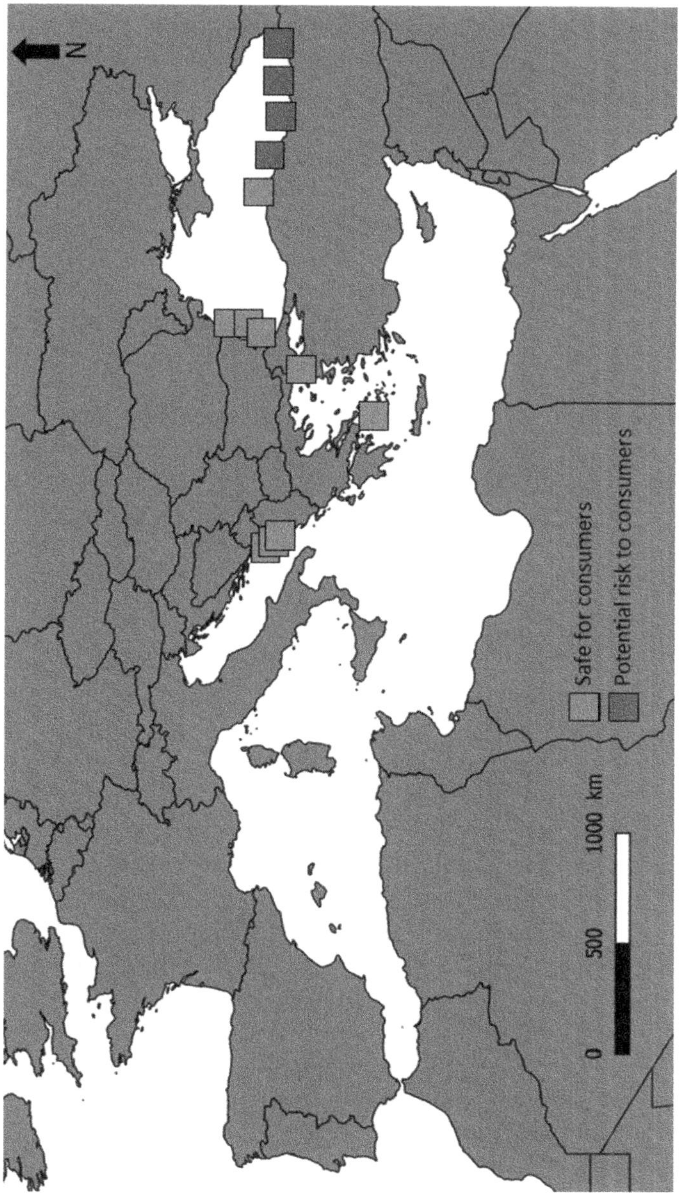

Figure 6.1. *Map showing human health risk assessment for consumers, based on trace elements, in Mytilus galloprovincialis from the Mediterranean Sea and Black Sea coasts, recently analyzed from various geographical locations [BAT 18, GED 17, JOV 15, KAL 12, PER 18, PEY 19, TAN 16, YIG 18, ZHE 18]*

As demonstrated, the TE concentrations in the whole edible soft tissues of mussels from different locations of the Mediterranean Sea and Black Sea, and the associated indices calculated (Figure 6.1, Table 6.3), show that the consumption of mussels from these areas can be considered as relatively safe for human health. Only one in nine studies has shown a potential risk to human health from dietary exposure.

The two main causes of TE contamination in the Mediterranean Sea and the Black Sea are industrialization and urbanization [BAT 18]. However, many other factors can play major roles, including specific geological features, intrinsic species-specific features, salinity, temperature, phytoplanktonic blooms and food availability [STA 12]. These external factors are known to influence TE bioaccumulation in organisms [SUS 10] and can partially explain our observed results.

Due to the increasing environmental and anthropogenic pressure on the Mediterranean Sea, and as this region will potentially be the largest producer of mussels in the future [STA 12], to assess human exposure to contaminants, regular monitoring of TE levels in mussels is required across the entire basin.

6.5. Perspectives in North Corsica ports

6.5.1. *Multibiomarker approach*

Contrary to the single biomarker approach commonly used in ecotoxicology studies, the multibiomarker approach allows us to assess the impacts of multiple contaminants, taking into account the environmental physico-chemical parameters, and allows us to differentiate the reversible effects from the irreversible ones [GAR 11]. This approach provides a "diagnosis of stress" that can be used to interpret an organism's level of physiological impairment [BRO 04]. It brings a better understanding and a more integrated view of the biological responses in environmental risk assessment [OZK 17, CRA 09]. *In situ* environmental studies, interactions between complex combinations of contaminants and environmental conditions require different biomarkers, especially when the level of contamination is not particularly high [CRA 09, AIT 06], like in North Corsica [GOB 09]. Moreover, several studies proved the efficiency of this kind of approach [VID 02, DAM 04, AIT 06, GAR 11, GUO 17].

Recently, the biomarker of exposure (bioaccumulation of TEs or organic pollutants) and the biomarker of effect (enzymes involved in the oxidative stress, immune system, energetic metabolism or neurotoxicity for example) were often chosen as effective biomarkers to evaluate different kinds of pollution [GUO 17, ULU 19, CAR 19, GHR 20]. However, due to the biological variability of organisms, it is always important to highlight useful biomarkers (responding to the exposition to pollution) from those that are less helpful for environmental assessment [CRA 09].

6.5.2. *The case of harbor areas*

Located in the coastal zone, harbor areas are subject to terrestrial discharges and anthropogenic contaminations to which the chronic contamination inherent in the port activity itself is added, with substances coming from a large number of diverse port activities, such as boat careening, recreational boating activities, fishing, maritime traffic, the use of antifouling paint, tourism, etc. Previous studies around the world have shown that levels of TEs [BUR 13, BRE 17, PIP 16], PAHs [MOS 16, LAI 15], butyltins [BUR 13, CAS 12] and organochlorine compounds [BEN 15, MAI 17] are particularly high in harbor water and sediments. The same was observed in Mediterranean coastal regions [LAF 07, LAF 08, AND 04, MAM 11]. In addition to chronic contamination, harbors can also be impacted by accidental oil spills [SUR 11, BRE 18]. The intense anthropogenic impact in harbors is reinforced by the low hydrodynamism of these areas that are generally sheltered, enclosed areas with a low flushing rate [FAS 12]. Overall, contamination in harbor areas is considered as a serious threat to the marine environment, and recreational activities could pose a subsequent risk to people's health [FAS 12, MAI 17, SUR 11, MAM 11].

6.5.3. *The QUAMPO project*

Based on all of this known knowledge and expertise, new projects continue to be introduced to the Mediterranean Sea because coastal waters, and especially harbor areas, need constant new information to maintain a good water quality and set up efficient and durable management of a changing environment. As an example, the QUAMPO (QUAlity of the marine environment in the Mediterranean POrt areas) project aims to evaluate the water quality of harbors in the north western Mediterranean Sea (around

Corsica) and defines tools of integrated biomonitoring for port authorities who seek ecomanagement accreditations. Focusing on three impacted locations (Saint-Florent, Calvi and Île Rousse harbors) and one reference site at the STARESO (*STAtion de REcherches Sous-marines et Océanographiques* de Calvi), QUAMPO will use a multibiomarker-based approach to evaluate the ecological status of the ecosystem in these specific locations. In addition, the environmental metabolomics approach will be applied using a non-targeted approach. This innovative method was shown to be a sensitive and powerful tool to screen xenobiotic metabolism in mussels and identify site-specific pollution [BON 17, FAS 12], but it is still scarcely applied in ecotoxicology. Finally, a model adapted to the water quality will be developed. Using the data generated by QUAMPO and other projects, a predictive approach will be developed, as is currently successfully done in coastal ecosystems with oysters [BAC 06, GRA 20, AHN 20]. QUAMPO is a collaboration between La Rochelle University and the STARESO, and is being funded for three years (2019–2022) by the European Maritime and Fisheries Fund (EMFF).

6.7. Acknowledgments

The authors would like to thank Mathilde Dabrowski and Claire Gillet for their help in gathering information and designing tables.

6.8. References

[ABD 19] ABDOU M., SCHÄFER J., HU R. *et al.*, "Platinum in sediments and mussels from the northwestern Mediterranean coast: Temporal and spatial aspects", *Chemosphere*, vol. 215, pp. 783–792, 2019.

[ADA 15] ADAM N., VERGAUWEN L., BLUST R. *et al.*, "Gene transcription patterns and energy reserves in *Daphnia magna* show no nanoparticle specific toxicity when exposed to ZnO and CuO nanoparticles", *Environmental Research*, vol. 138, pp. 82–92, 2015.

[AHN 20] AHN J.E., RONAN A.D., "Development of a model to assess coastal ecosystem health using oysters as the indicator species", *Estuarine, Coastal and Shelf Science*, vol. 233, p. 106528, 2020.

[AIT 06] AIT ALLA A., MOUNEYRAC C., DUROU C. *et al.*, "Tolerance and biomarkers as useful tools for assessing environmental quality in the Oued Souss estuary (Bay of Agadir, Morocco)", *Comparative Biochemistry and Physiology Part C: Toxicology and Pharmacology*, vol. 143, no. 1, pp. 23–29, 2006.

[AMI 06] AMIARD J.C., AMIARD-TRIQUET C., BARKA S. *et al.*, "Metallothioneins in aquatic invertebrates: Their role in metal detoxification and their use as biomarkers", *Aquatic Toxicology*, vol. 76, no. 2, pp. 160–202, 2006.

[AND 04] ANDRAL B., STANISIERE J.Y., SAUZADE D. *et al.*, "Monitoring chemical contamination levels in the Mediterranean based on the use of mussel caging", *Marine Pollution Bulletin*, vol. 49, no. 9, pp. 704–712, 2004.

[ASO 97] ASOKAN R., ARUMUGAM M., MULLAINADHAN P., "Activation of prophenoloxidase in the plasma and haemocytes of the marine mussel *Perna viridis* Linnaeus", *Developmental and Comparative Immunology*, vol. 21, no. 1, pp. 1–12, 1997.

[AUG 10] AUGIER H., *Guide des fonds marins de Méditerranée – écologie, flore, faune, plongée*, Delachaux et Niestlé, Paris, 2010.

[AZI 18a] AZIZI G., AKODAD M., BAGHOUR M. *et al.*, "The use of *Mytilus* sp. mussels as bioindicators of heavy metal pollution in the coastal environment. A review", *Journal of Materials and Environmental Science*, vol. 9, pp. 1170–1181, 2018.

[AZI 18b] AZIZI G., LAYACHI M., AKODAD M. *et al.*, "Seasonal variations of heavy metals content in mussels (*Mytilus galloprovincialis*) from Cala Iris offshore (Northern Morocco)", *Marine Pollution Bulletin*, vol. 137, pp. 688–694, 2018.

[BAC 06] BACHER C., GANGNERY A., "Use of dynamic energy budget and individual based models to simulate the dynamics of cultivated oyster populations", *Journal of Sea Research*, vol. 56, no. 2, pp. 140–155, 2006.

[BAE 05] BAEYENS W., LEERMAKERS M., GIETER M.D. *et al.*, "Overview of trace metal contamination in the Scheldt estuary and effect of regulatory measures", *Hydrobiologia*, vol. 540, no. 1, pp. 141–154, 2005.

[BAJ 19] BAJT O., RAMŠAK A., MILUN V. *et al.*, "Assessing chemical contamination in the coastal waters of the Adriatic Sea using active mussel biomonitoring with *Mytilus galloprovincialis*", *Marine Pollution Bulletin*, vol. 141, pp. 283–298, 2019.

[BAR 18] BARRICK A., MARION J.-M., PERREIN-ETTAJANI H. *et al.*, "Baseline levels of biochemical biomarkers in the endobenthic ragworm *Hediste diversicolor* as useful tools in biological monitoring of estuaries under anthropogenic pressure", *Marine Pollution Bulletin*, vol. 129, no. 1, pp. 81–85, 2018.

[BAT 17] BATTUELLO M., SARTOR R.M., BRIZIO P. *et al.*, "The influence of feeding strategies on trace element bioaccumulation in copepods (*Calanoida*)", *Ecological Indicators*, vol. 74, pp. 311–320, 2017.

[BAT 18] BAT L., ARICI E., ÖZTEKIN A., "Human health risk assessment of heavy metals in the Black Sea: Evaluating mussels", *Current World Environment*, vol. 13, no. 1, pp. 15–31, 2018.

[BEG 18] BEG M.U., BUTT S.A., AL-DUFAILEEJ S. *et al.*, "Biomarkers in fish as a measure of the state of marine environment of Kuwait", *Environmental Monitoring and Assessment*, vol. 190, no. 6, p. 325, 2018.

[BEN 11] BENEDICTO J., ANDRAL B., MARTÍNEZ-GÓMEZ C. *et al.*, "A large scale survey of trace metal levels in coastal waters of the Western Mediterranean basin using caged mussels (*Mytilus galloprovincialis*)", *Journal of Environmental Monitoring*, vol. 13, no. 5, pp. 1495–1505, 2011.

[BEN 15] BENALI I., BOUTIBA Z., MERABET A. *et al.*, "Integrated use of biomarkers and condition indices in mussels (*Mytilus galloprovincialis*) for monitoring pollution and development of biomarker index to assess the potential toxic of coastal sites", *Marine Pollution Bulletin*, vol. 95, no. 1, pp. 385–394, 2015.

[BER 00] BERGLUND O., LARSSON P., EWALD G. *et al.*, "Bioaccumulation and differential partitioning of polychlorinated biphenyls in freshwater, planktonic food webs", *Canadian Journal of Fisheries and Aquatic Sciences*, vol. 57, no. 6, pp. 1160–1168, 2000.

[BER 05] BERGÉ J.-P., BARNATHAN G., "Fatty acids from lipids of marine organisms: Molecular biodiversity, roles as biomarkers, biologically active compounds, and economical aspects", in ULBER, R., LE GAL, Y. (eds), *Marine Biotechnology I*, Springer, Berlin, Heidelberg, 2005.

[BER 12] BERLINE L., SIOKOU-FRANGOU I., MARASOVIĆ I. *et al.*, "Intercomparison of six Mediterranean zooplankton time series", *Progress in Oceanography*, vols 97–100, pp. 76–91, 2012.

[BET 80] BETHOUX J.P., "Mean water fluxes across sections in the Mediterranean Sea, evaluated on the basis of water and salt budgets and of observed salinities", *Oceanologica Acta*, vol. 3, no. 1, pp. 79–88, 1980.

[BET 99] BETHOUX J.P., GENTILI B., MORIN P. *et al.*, "The Mediterranean Sea: A miniature ocean for climatic and environmental studies and a key for the climatic functioning of the North Atlantic", *Progress in Oceanography*, vol. 44, no. 1, pp. 131–146, 1999.

[BEY 17] BEYER J., GREEN N.W., BROOKS S. *et al.*, "Blue mussels (*Mytilus edulis* sp.) as sentinel organisms in coastal pollution monitoring: A review", *Marine Environmental Research*, vol. 130, pp. 338–365, 2017.

[BIA 00] BIANCHI C.N., MORRI C., "Marine biodiversity of the Mediterranean Sea: Situation, problems and prospects for future research", *Marine Pollution Bulletin*, vol. 40, no. 5, pp. 367–376, 2000.

[BLA 86] BLANDIN P., "Bioindicateurs et diagnostic des systèmes écologiques", *Bulletin d'écologie*, vol. 17, no. 4, pp. 215–307, 1986.

[BOG 14] BOGDANOVIĆ T., UJEVIĆ I., SEDAK M. *et al.*, "As, Cd, Hg and Pb in four edible shellfish species from breeding and harvesting areas along the eastern Adriatic Coast, Croatia", *Food Chemistry*, vol. 146, pp. 197–203, 2014.

[BON 15] BONDARUK J., JANSON E., WYSOCKA M. *et al.*, "Identification of hazards for water environment in the upper Silesian Coal Basin caused by the discharge of salt mine water containing particularly harmful substances and radionuclides", *Journal of Sustainable Mining*, vol. 14, no. 4, pp. 179–187, 2015.

[BON 17] BONNEFILLE B., ARPIN-PONT L., GOMEZ E. *et al.*, "Metabolic profiling identification of metabolites formed in Mediterranean mussels (*Mytilus galloprovincialis*) after diclofenac exposure", *Science of the Total Environment*, vol. 583, pp. 257–268, 2017.

[BON 18] BONANNO G., ORLANDO-BONACA M., "Chemical elements in Mediterranean macroalgae. A review", *Ecotoxicology and Environmental Safety*, vol. 148, pp. 44–71, 2018.

[BOS 04] BOSC E., BRICAUD A., ANTOINE D., "Seasonal and interannual variability in algal biomass and primary production in the Mediterranean Sea, as derived from 4 years of SeaWiFS observations", *Global Biochemical Cycles*, vol. 18, no. 1, pp. 1–17, 2004.

[BRE 17] BREITWIESER M., VIRICEL A., CHURLAUD C. *et al.*, "First data on three bivalve species exposed to an intra-harbour polymetallic contamination (La Rochelle, France)", *Comparative Biochemistry and Physiology Part C: Toxicology and Pharmacology*, vol. 199, pp. 28–37, 2017.

[BRE 18] BREITWIESER M., THOMAS-GUYON H., HUET V. *et al.*, "Spatial and temporal impacts of the Skjervøy harbour diesel spill on native population of blue mussels: A sub-Arctic case study", *Ecotoxicology and Environmental Safety*, vol. 153, pp. 168–174, 2018.

[BRO 04] BROWN R.J., GALLOWAY T.S., LOWE D. *et al.*, "Differential sensitivity of three marine invertebrates to copper assessed using multiple biomarkers", *Aquatic Toxicology*, vol. 66, no. 3, pp. 267–278, 2004.

[BUR 08] BURGER J., "Assessment and management of risk to wildlife from cadmium", *Science of the Total Environment*, vol. 389, no. 1, pp. 37–45, 2008.

[BUR 13] BURUAEM L.M., DE CASTRO Í.B., HORTELLANI M.A. *et al.*, "Integrated quality assessment of sediments from harbour areas in Santos-São Vicente Estuarine System, Southern Brazil", *Estuarine, Coastal and Shelf Science*, vol. 130, pp. 179–189, 2013.

[CAH 72] CAHET G., FIALA M., JACQUES G. *et al.*, "Production primaire au niveau de la thermocline en zone néritique de Méditerranée Nord-Occidentale", *Marine Biology*, vol. 14, no. 1, pp. 32–40, 1972.

[CAP 15] CAPPELLO T., MAISANO M., GIANNETTO A. *et al.*, "Neurotoxicological effects on marine mussel *Mytilus galloprovincialis* caged at petrochemical contaminated areas (eastern Sicily, Italy): ^{1}H NMR and immunohistochemical assays", *Comparative Biochemistry and Physiology Part C: Toxicology and Pharmacology*, vol. 169, pp. 7–15, 2015.

[CAR 19] CARICATO R., GIORDANO M.E., SCHETTINO T. *et al.*, "Carbonic anhydrase integrated into a multimarker approach for the detection of the stress status induced by pollution exposure in *Mytilus galloprovincialis*: A field case study", *Science of the Total Environment*, vol. 690, pp. 140–150, 2019.

[CAS 12] CASTRO Í.B., ROSSATO M., FILLMANN G., "Imposex reduction and residual butyltin contamination in southern Brazilian harbors", *Environmental Toxicology and Chemistry in Latin America*, vol. 31, no. 5, pp. 947–954, 2012.

[CAS 15] CASTRITSI-CATHARIOS J., NEOFITOU N., VORLOOU A.A., "Comparison of heavy metal concentrations in fish samples from three fish farms (Eastern Mediterranean) utilizing antifouling paints", *Toxicological and Environmental Chemistry*, vol. 97, no. 1, pp. 116–123, 2015.

[CAT 93] CATSIKI V.A., PANAYOTIDIS P., "Copper, chromium and nickel in tissues of the Mediterranean seagrasses *Posidonia oceanica* and *Cymodocea nodosa* (Potamogetonaceae) from Greek coastal areas", *Chemosphere*, vol. 26, no. 5, pp. 963–978, 1993.

[CHO 11] CHOPRA A.K., SHARMA M.K., CHAMOLI S., "Bioaccumulation of organochlorine pesticides in aquatic system – An overview", *Environmental Monitoring and Assessment*, vol. 173, no. 1, pp. 905–916, 2011.

[COS 89] COSSA D., "A review of the use of *Mytilus* spp as quantitative indicators of cadmium and mercury contamination in coastal waters", *Oceanologica Acta*, vol. 12, no. 4, pp. 417–432, 1989.

[CRA 09] CRAVO A., LOPES B., SERAFIM Â. *et al.*, "A multibiomarker approach in *Mytilus galloprovincialis* to assess environmental quality", *Journal of Environmental Monitoring*, vol. 11, no. 9, pp. 1673–1686, 2009.

[DAG 14] D'AGATA A., CAPPELLO T., MAISANO M. *et al.*, "Cellular biomarkers in the mussel *Mytilus galloprovincialis* (Bivalvia: Mytilidae) from Lake Faro (Sicily, Italy)", *Italian Journal of Zoology*, vol. 81, no. 1, pp. 43–54, 2014.

[DAD 16] DADAR M., ADEL M., FERRANTE M. *et al.*, "Potential risk assessment of trace metals accumulation in food, water and edible tissue of rainbow trout (*Oncorhynchus mykiss*) farmed in Haraz River, northern Iran", *Toxin Reviews*, vol. 35, nos 3–4, pp. 141–146, 2016.

[DAI 11] DAILIANIS S., "Environmental impact of anthropogenic activities: The use of mussels as a reliable tool for monitoring marine pollution", in MCGEVIN, L.E. (ed.), *Mussels: Anatomy, Habitat and Environmental Impact*, Nova Sciences Publishers Inc., New York, 2011.

[DAM 04] DAMIENS G., HIS E., GNASSIA-BARELLI M. *et al.*, "Evaluation of biomarkers in oyster larvae in natural and polluted conditions", *Comparative Biochemistry and Physiology Part C: Toxicology and Pharmacology*, vol. 138, no. 2, pp. 121–128, 2004.

[DEM 06] DEMIRAK A., YILMAZ F., LEVENT TUNA A. *et al.*, "Heavy metals in water, sediment and tissues of *Leuciscus cephalus* from a stream in southwestern Turkey", *Chemosphere*, vol. 63, no. 9, pp. 1451–1458, 2006.

[DIB 15] DI BELLA G., POTORTÌ A.G., LO TURCO V. *et al.*, "Trace elements in *Thunnus thynnus* from Mediterranean Sea and benefit–risk assessment for consumers", *Food Additives and Contaminants: Part B*, vol. 8, no. 3, pp. 175–181, 2015.

[DIR 18] DIRRIGL F.J., BADAOUI Z., TAMEZ C. *et al.*, "Use of the sea hare (*Aplysia fasciata*) in marine pollution biomonitoring of harbors and bays", *Marine Pollution Bulletin*, vol. 129, no. 2, pp. 681–688, 2018.

[DUA 99] DUARTE C.M., AGUSTÍ S., KENNEDY H. *et al.*, "The Mediterranean climate as a template for Mediterranean marine ecosystems: The example of the northeast Spanish littoral", *Progress in Oceanography*, vol. 44, no. 1, pp. 245–270, 1999.

[DUA 11] DUARTE C.A., GIARRATANO E., AMIN O.A. *et al.*, "Heavy metal concentrations and biomarkers of oxidative stress in native mussels (*Mytilus edulis chilensis*) from Beagle Channel coast (Tierra del Fuego, Argentina)", *Marine Pollution Bulletin*, vol. 62, no. 8, pp. 1895–1904, 2011.

[DUP 20] DUPUIS H., NOUVEL G., THOMAS F., "Repères : pêche et aquaculture", in ABIS, S., BRUN, M. (ed.), *Le Déméter 2020*, IRIS editor, available at: https://www.iris-france.org/publications/le-demeter-2020/, 2020.

[DUR 11] DURRIEU DE MADRON X., GUIEU C., SEMPÉRÉ R. *et al.*, "Marine ecosystems' responses to climatic and anthropogenic forcings in the Mediterranean", *Progress in Oceanography*, vol. 91, no. 2, pp. 97–166, 2011.

[ERK 07] ERKAN N., ÖZDEN Ö., "Proximate composition and mineral contents in aqua cultured sea bass (*Dicentrarchus labrax*), sea bream (*Sparus aurata*) analyzed by ICP-MS", *Food Chemistry*, vol. 102, no. 3, pp. 721–725, 2007.

[EUR 06] EUROPEAN COMMISSION, Commission Regulation (EC) No 1881/2006 of 19 December 2006 setting maximum levels for certain contaminants in foodstuffs, Commission Regulation, European Commission, 2006.

[EUR 08a] EUROPEAN COMMISION, Directive 2008/56/EC of 17 June 2008 establishing a framework for community action in the field of marine environmental policy (Marine Strategy Framework Directive), Directive, European Commission, 2008.

[EUR 08b] EUROPEAN PARLIAMENT AND THE COUNCIL OF THE EUROPEAN UNION, Directive 2008/105/EC Environmental Quality Standards in the Field of Water Policy, Amending and Subsequently Repealing Council Directives 82/176/EEC, 83/513/EEC, 84/156/EEC, 84/491/EEC, 86/280/EEC and Amending Directive 2000/60/EC, Directive, European Commission, 2008.

[FAL 19] FALFUSHYNSKA H.I., WU F., YE F. *et al.*, "The effects of ZnO nanostructures of different morphology on bioenergetics and stress response biomarkers of the blue mussels *Mytilus edulis*", *Science of the Total Environment*, vol. 694, p. 133717, 2019.

[FAO 18] FAO, "The state of world fisheries and aquaculture 2018 – Meeting the sustainable development goals", *FAO*, Rome, 2018.

[FAR 16] FARRINGTON J.W., TRIPP B.W., TANABE S. *et al.*, "Edward D. Goldberg's proposal of 'the Mussel Watch': Reflections after 40 years", *Marine Pollution Bulletin*, vol. 110, no. 1, pp. 501–510, 2016.

[FAS 12] FASULO S., IACONO F., CAPPELLO T. *et al.*, "Metabolomic investigation of *Mytilus galloprovincialis* (Lamarck 1819) caged in aquatic environments", *Ecotoxicology and Environmental Safety*, vol. 84, pp. 139–146, 2012.

[FER 08] FERNÁNDEZ DE PUELLES M.L., MOLINERO J.C., "Decadal changes in hydrographic and ecological time-series in the Balearic Sea (western Mediterranean), identifying links between climate and zooplankton", *ICES Journal of Marine Science*, vol. 65, no. 3, pp. 311–317, 2008.

[FER 10] FERNÁNDEZ B., CAMPILLO J.A., MARTÍNEZ-GÓMEZ C. *et al.*, "Antioxidant responses in gills of mussel (*Mytilus galloprovincialis*) as biomarkers of environmental stress along the Spanish Mediterranean coast", *Aquatic Toxicology*, vol. 99, no. 2, pp. 186–197, 2010.

[FIT 93] FITZPATRICK P.J., SHEEHAN D., "Separation of multiple forms of glutathione S-transferase from the Blue Mussel, *Mytilus edulis*", *Xenobiotica*, vol. 23, no. 8, pp. 851–861, 1993.

[FOK 13] FOKINA N.N., RUOKOLAINEN T.R., NEMOVA N.N. *et al.*, "Changes of blue mussels *Mytilus edulis* L. lipid composition under cadmium and copper toxic effect", *Biological Trace Element Research*, vol. 154, no. 2, pp. 217–225, 2013.

[FRA 05] FRAGA C., "Relevance, essentiality and toxicity of trace elements in human health", *Molecular Aspects of Medicine*, vol. 26, pp. 235–44, 2005.

[FUL 20] FULLGRABE L., GROSJEAN P., GOBERT S. *et al.*, "Zooplankton dynamics in a changing environment: A 13-year survey in the northwestern Mediterranean Sea", *Marine Environmental Research*, p. 104962, 2020.

[GAG 14] GAGNÉ F., "Oxidative stress", in GAGNÉ, F. (ed.), *Biochemical Ecotoxicology*, Academic Press, Oxford, 2014.

[GAL 11] GALGANI F., MARTÍNEZ-GÓMEZ C., GIOVANARDI F. *et al.*, "Assessment of polycyclic aromatic hydrocarbon concentrations in mussels (*Mytilus galloprovincialis*) from the Western basin of the Mediterranean Sea", *Environmental Monitoring and Assessment*, vol. 172, no. 1, pp. 301–317, 2011.

[GAR 10] GARRIC J., MORIN S., VINCENT-HUBERT F., "Les biomarqueurs en écotoxicologie : définition, intérêt, limite, usage", *Sciences eaux et territoires*, vol. 1, pp. 12–17, 2010.

[GAR 11] GARMENDIA L., SOTO M., ORTIZ-ZARRAGOITIA M. *et al.*, "Application of a battery of biomarkers in mussel digestive gland to assess long-term effects of the Prestige oil spill in Galicia and Bay of Biscay: Correlation and multivariate analysis", *Journal of Environmental Monitoring*, vol. 13, no. 4, pp. 933–942, 2011.

[GED 17] GEDIK K., "Bioaccessibility of Cd, Cr, Cu, Mn, Ni, Pb, and Zn in Mediterranean mussel (*Mytilus galloprovincialis* Lamarck, 1819) along the southeastern Black Sea coast", *Human and Ecological Risk Assessment: An International Journal*, pp. 1–13, 2017.

[GHR 20] GHRIBI F., RICHIR J., BEJAOUI S. *et al.*, "Trace elements and oxidative stress in the Ark shell *Arca noae* from a Mediterranean coastal lagoon (Bizerte lagoon, Tunisia): Are there health risks associated with their consumption?", *Environmental Science and Pollution Research*, pp. 1–17, 2020.

[GOB 09] GOBERT S., SARTORETTO S., RICO-RAIMONDINO V. *et al.*, "Assessment of the ecological status of Mediterranean French coastal waters as required by the Water Framework Directive using the *Posidonia oceanica* Rapid Easy Index: PREI", *Marine Pollution Bulletin*, vol. 58, no. 11, pp. 1727–1733, 2009.

[GOL 75] GOLDBERG E.D., "The mussel watch – A first step in global marine monitoring", *Marine Pollution Bulletin*, vol. 6, no. 7, p. 111, 1975.

[GOL 80] GOLDBERG E., *The International Mussel Watch: Report of a Workshop*, The National Academies Press, Washington, DC, 1980.

[GRA 20] GRAHAM P., BRUNDU G., SCOLAMACCHIA M. *et al.*, "Improving Pacific oyster (*Crassostrea gigas*, Thunberg, 1793) production in Mediterranean coastal lagoons: Validation of the growth model 'ShellSIM' on traditional and novel farming methods", *Aquaculture*, vol. 516, p. 734612, 2020.

[GUO 17] GUO R., PAN L., JI R., "A multi-biomarker approach in scallop *Chlamys farreri* to assess the impact of contaminants in Qingdao coastal area of China", *Ecotoxicology and Environmental Safety*, vol. 142, pp. 399–409, 2017.

[HAB 74] HABIG W.H., PABST M.J., JAKOBY W.B., "Glutathione S-transferases. The first enzymatic step in mercapturic acid formation", *The Journal of Biological Chemistry*, vol. 249, no. 22, pp. 7130–7139, 1974.

[HAL 99] HALLIWELL B., GUTTERIDGE J.M.C., *Free Radicals in Biology and Medicine*, Oxford University Press, Oxford, 1999.

[HOL 18] HOLON F., MARRE G., PARRAVICINI V. *et al.*, "A predictive model based on multiple coastal anthropogenic pressures explains the degradation status of a marine ecosystem: Implications for management and conservation", *Biological Conservation*, vol. 222, pp. 125–135, 2018.

[IFR 74] IFREMER, Réseau d'Observation de la Contamination CHimique du littoral (ROCCH), available at: https://wwz.ifremer.fr/littoral/Reseaux-de-surveillance/Environnement/ROCCH-Reseau-d-Observation-de-la-Contamination-CHimique-du-littoral, 1974.

[INE 17] INERIS, Portail substances chimiques, available at: http://www.ineris.fr/substances/fr/, 2017.

[ISL 04] ISLAM S.M., TANAKA M., "Impacts of pollution on coastal and marine ecosystems including coastal and marine fisheries and approach for management: A review and synthesis", *Marine Pollution Bulletin*, vol. 48, no. 7, pp. 624–649, 2004.

[JEC 11] JECFA, Safety evaluation of certain food additives and contaminants, 73rd meeting of the Joint FAO/WHO Expert Committee on Food Additives, WHO food additives series no. 64, World Health Organization, Geneva, 2011.

[JI 19] JI R., PAN L., GUO R. *et al.*, "Using multi-integrated biomarker indexes approach to assess marine quality and health status of marine organism: A case study of *Ruditapes philippinarum* in Laizhou Bay, China", *Environmental Science and Pollution Research*, vol. 26, no. 10, pp. 9916–9930, 2019.

[JOH 06] JOHANSSON I., HÉAS-MOISAN K., GUIOT N. *et al.*, "Polybrominated diphenyl ethers (PBDEs) in mussels from selected French coastal sites: 1981–2003", *Chemosphere*, vol. 64, no. 2, pp. 296–305, 2006.

[JOV 15] JOVIC M., ONJIA A., TANASKOVSKI B. *et al.*, "Human exposure to 17 elements through mussels consumption", *VII International Conference: Water & Fish*, Belgrade, Serbia, June 10–12, 2015.

[KAA 99] KAAYA A., NAJIMI S., RIBERA D. *et al.*, "Characterization of Glutathione S-Transferases (GST) activities in *Perna perna* and *Mytilus galloprovincialis* used as a biomarker of pollution in the Agadir marine Bay (South of Morocco)", *Bulletin of Environmental Contamination and Toxicology*, vol. 62, no. 5, pp. 623–629, 1999.

[KAL 12] KALOGEROPOULOS N., KARAVOLTSOS S., SAKELLARI A. *et al.*, "Heavy metals in raw, fried and grilled Mediterranean finfish and shellfish", *Food and Chemical Toxicology*, vol. 50, no. 10, pp. 3702–3708, 2012.

[KIM 17] KIM M., LI L. Y., GORGY T. *et al.*, "Review of contamination of sewage sludge and amended soils by polybrominated diphenyl ethers based on meta-analysis", *Environmental Pollution*, vol. 220, pp. 753–765, 2017.

[KOU 98] KOUWENBERG J., Shift in copepod populations and long-term changes in the northwestern Mediterranean: An overview, UNESCO Intergovernmental Oceanographic Commission Workshop Report, pp. 203–213, 1998.

[KRI 18] KRISHNAKUMAR P.K., QURBAN M.A., SASIKUMAR G., "Biomonitoring of trace metals in the coastal waters using bivalve molluscs", in SALEH, H.E.-D., EL-ADHAM, E. (eds), *Trace Elements – Human Health and Environment*, IntechOpen, London, 2018.

[KWO 09] KWONG R.W.M., YU P.K.N., LAM P.K.S. *et al.*, "Biokinetics and biotransformation of DDTs in the marine green mussels *Perna viridis*", *Aquatic Toxicology*, vol. 93, no. 4, pp. 196–204, 2009.

[LAF 07] LAFABRIE C., PERGENT G., KANTIN R. *et al.*, "Trace metals assessment in water, sediment, mussel and seagrass species – Validation of the use of *Posidonia oceanica* as a metal biomonitor", *Chemosphere*, vol. 68, no. 11, pp. 2033–2039, 2007.

[LAF 08] LAFABRIE C., PERGENT-MARTINI C., PERGENT G., "Metal contamination of *Posidonia oceanica* meadows along the Corsican coastline (Mediterranean)", *Environmental Pollution*, vol. 151, no. 1, pp. 262–268, 2008.

[LAI 15] LAITANO M.V., CASTRO Í.B., COSTA P.G. *et al.*, "Butyltin and PAH contamination of Mar del Plata port (Argentina) sediments and their influence on adjacent coastal regions", *Bulletin of Environmental Contamination and Toxicology*, vol. 95, no. 4, pp. 513–520, 2015.

[LEC 19] LE CROIZIER G., LACROIX C., ARTIGAUD S. *et al.*, "Metal subcellular partitioning determines excretion pathways and sensitivity to cadmium toxicity in two marine fish species", *Chemosphere*, vol. 217, pp. 754–762, 2019.

[LEF 03] LE FLOCH S., GUYOMARCH J., MERLIN F. *et al.*, "Effects of oil and bioremediation on mussel (*Mytilus edulis* L.) growth in mudflats", *Environmental Technology*, vol. 24, no. 10, pp. 1211–1219, 2003.

[LEI 13] LEI B., CHEN L., HAO Y. *et al.*, "Trace elements in animal-based food from Shanghai markets and associated human daily intake and uptake estimation considering bioaccessibility", *Ecotoxicology and Environmental Safety*, vol. 96, pp. 160–167, 2013.

[LEÓ 13] LEÓN V.M., MARTÍNEZ-GÓMEZ C., GARCÍA I. *et al.*, "Spatial distribution and temporal trends of polycyclic aromatic hydrocarbons in *Mytilus galloprovincialis* from the Iberian Mediterranean coast", *Environmental Monitoring and Assessment*, vol. 185, no. 2, pp. 1055–1070, 2013.

[LIO 03] LIONETTO M.G., CARICATO R., GIORDANO M.E. *et al.*, "Integrated use of biomarkers (acetylcholinesterase and antioxidant enzymes activities) in *Mytilus galloprovincialis* and *Mullus barbatus* in an Italian coastal marine area", *Marine Pollution Bulletin*, vol. 46, no. 3, pp. 324–330, 2003.

[LIO 13] LIONETTO M.G., CARICATO R., CALISI A. *et al.*, "Acetylcholinesterase as a biomarker in environmental and occupational medicine: New insights and future perspectives", *BioMed Research International*, vol. 2013, pp. 321213, 2013.

[LIV 91] LIVINGSTONE D.R., "Origins and evolution of pathways of anaerobic metabolism in the animal kingdom", *American Zoologist*, vol. 31, no. 3, pp. 522–534, 1991.

[LIV 01] LIVINGSTONE D.R., "Contaminant-stimulated reactive oxygen species production and oxidative damage in aquatic organisms", *Marine Pollution Bulletin*, vol. 42, no. 8, pp. 656–666, 2001.

[LOA 20] LOAIZA I., PILLET M., DE BOECK G. *et al.*, "Peruvian scallop *Argopecten purpuratus*: From a key aquaculture species to a promising biondicator species", *Chemosphere*, vol. 239, p. 124767, 2020.

[LOW 00] LOWE D.M., FOSSATO V.U., "The influence of environmental contaminants on lysosomal activity in the digestive cells of mussels (*Mytilus galloprovincialis*) from the Venice Lagoon", *Aquatic Toxicology*, vol. 48, nos 2–3, pp. 75–85, 2000.

[LUN 11] LUNA-ACOSTA A., SAULNIER D., POMMIER M. *et al.*, "First evidence of a potential antibacterial activity involving a laccase-type enzyme of the phenoloxidase system in Pacific oyster *Crassostrea gigas* haemocytes", *Fish and Shellfish Immunology*, vol. 31, no. 6, pp. 795–800, 2011.

[LUN 15] LUNA-ACOSTA A., BUSTAMANTE P., BUDZINSKI H. *et al.*, "Persistent organic pollutants in a marine bivalve on the Marennes-Oléron Bay and the Gironde Estuary (French Atlantic Coast) – Part 2: Potential biological effects", *Science of the Total Environment*, vol. 514, pp. 511–522, 2015.

[LUN 17] LUNA-ACOSTA A., BREITWIESER M., RENAULT T. *et al.*, "Recent findings on phenoloxidases in bivalves", *Marine Pollution Bulletin*, vol. 122, nos 1–2, pp. 5–16, 2017.

[LUS 11] LUSHCHAK V.I., "Environmentally induced oxidative stress in aquatic animals", *Aquatic Toxicology*, vol. 101, pp. 13–30, 2011.

[MAI 17] MAISANO M., CAPPELLO T., NATALOTTO A. *et al.*, "Effects of petrochemical contamination on caged marine mussels using a multi-biomarker approach: Histological changes, neurotoxicity and hypoxic stress", *Marine Environmental Research*, vol. 128, pp. 114–123, 2017.

[MAM 11] MAMINDY-PAJANY Y., HAMER B., ROMÉO M. *et al.*, "The toxicity of composted sediments from Mediterranean ports evaluated by several bioassays", *Chemosphere*, vol. 82, no. 3, pp. 362–369, 2011.

[MAR 05] MARASOVIĆ I., NINČEVIĆ Ž., KUŠPILIĆ G. *et al.*, "Long-term changes of basic biological and chemical parameters at two stations in the middle Adriatic", *Journal of Sea Research*, vol. 54, no. 1, pp. 3–14, 2005.

[MAR 17] MARTÍNEZ-GÓMEZ C., ROBINSON C.D., BURGEOT T. *et al.*, "Biomarkers of general stress in mussels as common indicators for marine biomonitoring programmes in Europe: The ICON experience", *Marine Environmental Research*, vol. 124, pp. 70–80, 2017.

[MAR 18] MARENGO M., DURIEUX E.D.H., TERNENGO S. *et al.*, "Comparison of elemental composition in two wild and cultured marine fish and potential risks to human health", *Ecotoxicology and Environmental Safety*, vol. 158, pp. 204–212, 2018.

[MAT 05] MATOZZO V., TOMEI A., MARIN M.G., "Acetylcholinesterase as a biomarker of exposure to neurotoxic compounds in the clam *Tapes philippinarum* from the Lagoon of Venice", *Marine Pollution Bulletin*, vol. 50, no. 12, pp. 1686–1693, 2005.

[MAZ 03] MAZZOCCHI M.G., NERVEGNA D., D'ELIA G. *et al.*, "Spring mesozooplankton communities in the epipelagic Ionian Sea in relation to the Eastern Mediterranean Transient", *Journal of Geophysical Research*, vol. 108, no. C9, 2003.

[MAZ 07] MAZZOCCHI M.G., CHRISTOU E.D., CAPUA I.D. *et al.*, "Temporal variability of *Centropages typicus* in the Mediterranean Sea over seasonal-to-decadal scales", *Progress in Oceanography*, vol. 72, no. 2, pp. 214–232, 2007.

[MCC 69] McCord J.M., Fridovich I., "Superoxide dismutase. An enzymic function for erythrocuprein (hemocuprein)", *Journal of Biological Chemistry*, vol. 244, no. 22, pp. 6049–6055, 1969.

[MED 12] Medeiros R.J., Dos Santos L.M.G., Freire A.S. *et al.*, "Determination of inorganic trace elements in edible marine fish from Rio de Janeiro State, Brazil", *Food Control*, vol. 23, no. 2, pp. 535–541, 2012.

[MEG 19] Megson D., Benoit N.B., Sandau C.D. *et al.*, "Evaluation of the effectiveness of different indicator PCBs to estimating total PCB concentrations in environmental investigations", *Chemosphere*, vol. 237, p. 124429, 2019.

[MIL 16] Milun V., Grgas D., Dragičević T.L., "Assessment of PCB and chlorinated pesticide accumulation in mussels at Kaštela Bay (Eastern Adriatic)", *Science of the Total Environment*, vol. 562, pp. 115–127, 2016.

[MON 07] Monserrat J.M., Martínez P.E., Geracitano L.A. *et al.*, "Pollution biomarkers in estuarine animals: Critical review and new perspectives", *Comparative Biochemistry and Physiology Part C: Toxicology & Pharmacology*, vol. 146, no. 1, pp. 221–234, 2007.

[MON 17] Monneret C., "What is an endocrine disruptor?", *Comptes rendus biologies*, vol. 340, no. 9, pp. 403–405, 2017.

[MOO 04] Moore M., Lowe D., Koehler A., "Biological effects of contaminants: Measurement of lysosomal membrane stability", *ICES Techniques in Marine Environmental Sciences*, no. 36, International Council for the Exploration of the Sea, Copenhagen, Denmark, November 2004.

[MOS 16] Moschino V., Del Negro P., De Vittor C. *et al.*, "Biomonitoring of a polluted coastal area (Bay of Muggia, Northern Adriatic Sea): A five-year study using transplanted mussels", *Ecotoxicology and Environmental Safety*, vol. 128, pp. 1–10, 2016.

[NAS 02] Nasci C., Nesto N., Monteduro R.A. *et al.*, "Field application of biochemical markers and a physiological index in the mussel, *Mytilus galloprovincialis*: Transplantation and biomonitoring studies in the lagoon of Venice (NE Italy)", *Marine Environmental Research*, vol. 54, no. 3, pp. 811–816, 2002.

[ORA 18] Orani A.M., Barats A., Vassileva E. *et al.*, "Marine sponges as a powerful tool for trace elements biomonitoring studies in coastal environment", *Marine Pollution Bulletin*, vol. 131, pp. 633–645, 2018.

[ÖZD 10] Özden Ö., Ulusoy Ş., Erkan N., "Study on the behavior of the trace metal and macro minerals in *Mytilus galloprovincialis* as a bioindicator species: The case of Marmara Sea, Turkey", *Journal für Verbraucherschutz und Lebensmittelsicherheit*, vol. 5, no. 3, pp. 407–412, 2010.

[OZK 17] OZKAN D., DAGDEVIREN M., KATALAY S. *et al.*, "Multi-biomarker responses after exposure to pollution in the Mediterranean mussels (*Mytilus galloprovincialis* L.) in the Aegean coast of Turkey", *Bulletin of Environmental Contamination and Toxicology*, vol. 98, no. 1, pp. 46–52, 2017.

[PAM 05] PAMPANIN D.M., CAMUS L., GOMIERO A. *et al.*, "Susceptibility to oxidative stress of mussels (*Mytilus galloprovincialis*) in the Venice Lagoon (Italy)", *Marine Pollution Bulletin*, vol. 50, no. 12, pp. 1548–1557, 2005.

[PAR 20] PARRA-LUNA M., MARTÍN-POZO L., HIDALGO F. *et al.*, "Common sea urchin (*Paracentrotus lividus*) and sea cucumber of the genus *Holothuria* as bioindicators of pollution in the study of chemical contaminants in aquatic media. A revision", *Ecological Indicators*, vol. 113, p. 106185, 2020.

[PEL 09] PELLEGRINI M., BALDARI C.T., "Apoptosis and oxidative stress-related diseases: The p66Shc connection", *Current Molecular Medicine*, vol. 9, no. 3, pp. 392–398, 2009.

[PER 98] PERGENT-MARTINI C., "*Posidonia oceanica*: A biological indicator of past and present mercury contamination in the Mediterranean Sea", *Marine Environmental Research*, vol. 45, no. 2, pp. 101–111, 1998.

[PER 00] PERGENT-MARTINI C., PERGENT G., "Marine phanerogams as a tool in the evaluation of marine trace-metal contamination: An example from the Mediterranean", *International Journal of Environment and Pollution*, vol. 13, nos 1–6, pp. 126–147, 2000.

[PER 18] PEROŠEVIĆ A., JOKSIMOVIĆ D., ĐUROVIĆ D. *et al.*, "Human exposure to trace elements via consumption of mussels *Mytilus galloprovincialis* from Boka Kotorska Bay, Montenegro", *Journal of Trace Elements in Medicine and Biology*, vol. 50, pp. 554–559, 2018.

[PER 19] PERIĆ L., BURIĆ P., "The effect of copper and chlorpyrifos co-exposure on biomarkers in the marine mussel *Mytilus galloprovincialis*", *Chemosphere*, vol. 225, pp. 126–134, 2019.

[PEY 19] PEYCHEVA K., PANAYOTOVA V., MERDZHANOVA A. *et al.*, "Estimation of THQ and potential health risk for metals by comsumption of some black sea marine fishes and mussels in Bulgaria", *Bulgarian Chemical Communications*, vol. 51, no. D, pp. 241–246, 2019.

[PHI 76] PHILLIPS D.J.H., "The common mussel *Mytilus edulis* as an indicator of pollution by zinc, cadmium, lead and copper I. Effects of environmental variables on uptake of metals", *Marine Biology*, vol. 38, no. 1, pp. 59–69, 1976.

[PIC 14] PICOT-GROZ M., MARTINEZ BUENO M.J., ROSAIN D. *et al.*, "Detection of emerging contaminants (UV filters, UV stabilizers and musks) in marine mussels from Portuguese coast by QuEChERS extraction and GC–MS/MS", *Science of the Total Environment*, vol. 493, pp. 162–169, 2014.

[PIN 15] PINI J.M., RICHIR J., WATSON G.J., "Metal bioavailability and bioaccumulation in the polychaete *Nereis (Alitta) virens* (Sars): The effects of site-specific sediment characteristics", *Marine Pollution Bulletin*, vol. 95, no. 2, pp. 565–575, 2015.

[PIP 16] PIPPY B.A., KIDD K.A., MUNKITTRICK K.R. *et al.*, "Use of the Atlantic nut clam (*Nucula proxima*) and catworm (*Nephtys incisa*) in a sentinel species approach for monitoring the health of Bay of Fundy estuaries", *Marine Pollution Bulletin*, vol. 106, no. 1, pp. 225–235, 2016.

[PYT 08] PYTHAROPOULOU S., SAZAKLI E., GRINTZALIS K. *et al.*, "Translational responses of *Mytilus galloprovincialis* to environmental pollution: Integrating the responses to oxidative stress and other biomarker responses into a general stress index", *Aquatic Toxicology*, vol. 89, pp. 18–27, 2008.

[RAD 88] RADI A.A.R., MATKOVICS B., "Effects of metal ions on the antioxidant enzyme activities, protein contents and lipid peroxidation of carp tissues", *Comparative Biochemistry and Physiology Part C: Comparative Pharmacology*, vol. 90, pp. 69–72, 1988.

[RAI 93] RAINBOW P.S., PHILLIPS D.J.H., "Cosmopolitan biomonitors of trace metals", *Marine Pollution Bulletin*, vol. 26, no. 11, pp. 593–601, 1993.

[RAI 02] RAINBOW P.S., "Trace metal concentrations in aquatic invertebrates: Why and so what?", *Environmental Pollution*, vol. 120, pp. 497–507, 2002.

[RAZ 91] RAZOULS C., DURAND J., "Inventaire des copépodes planctoniques méditerranéens", *Vie et milieu*, vol. 41, no. 1, pp. 73–77, 1991.

[RED 17] REDMOND K.J., BERRY M., PAMPANIN D.M. *et al.*, "Valve gape behaviour of mussels (*Mytilus edulis*) exposed to dispersed crude oil as an environmental monitoring endpoint", *Marine Pollution Bulletin*, vol. 117, no. 1, pp. 330–339, 2017.

[REG 93] REGOLI F., ORLANDO E., "*Mytilus galloprovincialis* as a bioindicator of lead pollution: Biological variables and cellular responses", *Science of the Total Environment*, vol. 134, pp. 1283–1292, 1993.

[REG 95] REGOLI F., PRINCIPATO G., "Glutathione, glutathione-dependent and antioxidant enzymes in mussel, *Mytilus galloprovincialis*, exposed to metals under field and laboratory conditions: Implications for the use of biochemical biomarkers", *Aquatic Toxicology*, vol. 31, no. 2, pp. 143–164, 1995.

[REN 13] REN X.-M., GUO L.-H., GAO Y. *et al.*, "Hydroxylated polybrominated diphenyl ethers exhibit different activities on thyroid hormone receptors depending on their degree of bromination", *Toxicology and Applied Pharmacology*, vol. 268, no. 3, pp. 256–263, 2013.

[RIB 04] RIBERA D'ALCALA M., CONVERSANO F., CORATO F. *et al.*, "Seasonal patterns in plankton communities in a pluriannual time series at a coastal Mediterranean site (Gulf of Naples): An attempt to discern recurrences and trends", *Scienca Marina*, vol. 68, pp. 65–83, 2004.

[RIC 12] RICHIR J., Coastal pollution of the Mediterranean and extension of its biomonitoring to trace elements of emerging concern, PhD thesis, University of Liège, Liège, 2012.

[RIC 14] RICHIR J., GOBERT S., "A reassessment of the use of *Posidonia oceanica* and *Mytilus galloprovincialis* to biomonitor the coastal pollution of trace elements: New tools and tips", *Marine Pollution Bulletin*, vol. 89, nos 1–2, pp. 390– 406, 2014.

[RIC 16a] RICHIR J., GOBERT S., "Trace elements in marine environments: Occurrence, threats and monitoring with special focus on the coastal Mediterranean", *Journal of Environmental and Analytical Toxicology*, vol. 6, no. 1, 2016.

[RIC 16b] RICHIR J., GOBERT S., "An ecophysiological discussion of trace element bioaccumulation in cultured *Mytilus galloprovincialis*", *Belgian Journal of Zoology*, vol. 145, no. 1, pp. 53–61, 2016.

[ROB 03] ROBILLARD S., BEAUCHAMP G., LAULIER M., "The role of abiotic factors and pesticide levels on enzymatic activity in the freshwater mussel *Anodonta cygnea* at three different exposure sites", *Comparative Biochemistry and Physiology Part C: Toxicology and Pharmacology*, vol. 135, no. 1, pp. 49–59, 2003.

[ROC 14] ROCCHETTA I., PASQUEVICH M.Y., HERAS H. *et al.*, "Effects of sewage discharges on lipid and fatty acid composition of the Patagonian bivalve *Diplodon chilensis*", *Marine Pollution Bulletin*, vol. 79, no. 1, pp. 211–219, 2014.

[ROM 97] ROMÉO M., GNASSIA-BARELLI M., "Effect of heavy metals on lipid peroxidation in the Mediterranean clam *Ruditapes decussatus*", *Comparative Biochemistry and Physiology Part C: Pharmacology, Toxicology and Endocrinology*, vol. 118, no. 1, pp. 33–37, 1997.

[ROM 03] ROMÉO M., HOARAU P., GARELLO G. *et al.*, "Mussel transplantation and biomarkers as useful tools for assessing water quality in the NW Mediterranean", *Environmental Pollution*, vol. 122, no. 3, pp. 369–378, 2003.

[ROS 08] ROSSI N., JAMET J.-L., "*In situ* heavy metals (copper, lead and cadmium) in different plankton compartments and suspended particulate matter in two coupled Mediterranean coastal ecosystems (Toulon Bay, France)", *Marine Pollution Bulletin*, vol. 56, no. 11, pp. 1862–1870, 2008.

[SCH 09] SCHIPPER C., Assessment of effects of chemical contaminants in dredged material on marine ecosystems and human health, PhD thesis, Wageningen University, 2009.

[SER 16] SERRANITO B., AUBERT A., STEMMANN L. *et al.*, "Proposition of indicators of anthropogenic pressure in the Bay of Toulon (Mediterranean Sea) based on zooplankton time-series", *Continental Shelf Research*, vol. 121, pp. 3–12, 2016.

[SEV 11] SEVCIKOVA M., MODRA H., SLANINOVA A. *et al.*, "Metals as a cause of oxidative stress in fish: A review", *Veterinarni Medicina*, vol. 56, no. 11, pp. 537–546, 2011.

[SIO 91] SIOKOU-FRANGOU I., PAPATHANASSIOU E., "Differentiation of zooplankton populations in a polluted area", *Marine Ecology Progress Series*, vol. 76, p. 41–51, 1991.

[SÖD 98] SÖDERHÄLL K., CERENIUS L., "Role of the prophenoloxidase-activating system in invertebrate immunity", *Current Opinion in Immunology*, vol. 10, no. 1, pp. 23–28, 1998.

[SOL 18] SOLÉ M., RIVERA-INGRAHAM G., FREITAS R., "The use of carboxylesterases as biomarkers of pesticide exposure in bivalves: A methodological approach", *Comparative Biochemistry and Physiology Part C: Toxicology & Pharmacology*, vol. 212, pp. 18–24, 2018.

[SQU 16] SQUADRONE S., BRIZIO P., STELLA C. *et al.*, "Presence of trace metals in aquaculture marine ecosystems of the northwestern Mediterranean Sea (Italy)", *Environmental Pollution*, vol. 215, pp. 77–83, 2016.

[SRI 98] SRIDEVI B., REDDY K.V., REDDY S.L.N., "Effect of trivalent and hexavalent chromium on antioxidant enzyme activities and lipid peroxidation in a freshwater field crab, *Barytelphusa guerini*", *Bulletin of Environmental Contamination and Toxicology*, vol. 61, no. 3, pp. 384–390, 1998.

[STA 12] STANKOVIC S., JOVIC M., "Health risks of heavy metals in the Mediterranean mussels as seafood", *Environmental Chemistry Letters*, vol. 10, no. 2, pp. 119–130, 2012.

[STO 95] STOHS S.J., BAGCHI D., "Oxidative mechanisms in the toxicity of metal ions", *Free Radical Biology and Medicine*, vol. 18, no. 2, pp. 321–336, 1995.

[SUR 11] SUREDA A., BOX A., TEJADA S. *et al.*, "Biochemical responses of *Mytilus galloprovincialis* as biomarkers of acute environmental pollution caused by the Don Pedro oil spill (Eivissa Island, Spain)", *Aquatic Toxicology*, vol. 101, no. 3, pp. 540–549, 2011.

[SUS 10] SUSENO H., HUDIYONO S., BUDIAWAN B. *et al.*, "Effects of concentration, body size and food type on the bioaccumulation of Hg in farmed tilapia *Oreochromis mossambicus*", *Australian Journal of Basic and Applied Sciences*, vol. 4, pp. 792–799, 2010.

[SVI 15] SVINGEN T., VINGGAARD A.M., "The risk of chemical cocktail effects and how to deal with the issue", *Journal of Epidemiology and Community Health*, vol. 0, pp. 1–2, 2015.

[TAN 16] TANASKOVSKI B., JOVIĆ M., MANDIĆ M. *et al.*, "Elemental analysis of mussels and possible health risks arising from their consumption as a food: The case of Boka Kotorska Bay, Adriatic Sea", *Ecotoxicology and Environmental Safety*, vol. 130, pp. 65–73, 2016.

[TAP 82] TAPPEL M.E., CHAUDIERE J., TAPPEL A.L., "Glutathione peroxidase activities of animal tissues", *Comparative Biochemistry and Physiology Part B: Comparative Biochemistry*, vol. 73, no. 4, pp. 945–949, 1982.

[TET 08] TETT P., CARREIRA C., MILLS D.K. *et al.*, "Use of a Phytoplankton Community Index to assess the health of coastal waters", *ICES Journal of Marine Science*, vol. 65, no. 8, pp. 1475–1482, 2008.

[TSA 16] TSANGARIS C., MOSCHINO V., STROGYLOUDI E. *et al.*, "Biochemical biomarker responses to pollution in selected sentinel organisms across the Eastern Mediterranean and the Black Sea", *Environmental Science and Pollution Research*, vol. 23, no. 2, pp. 1789–1804, 2016.

[TUR 99] TURLEY C.M., "The changing Mediterranean Sea – A sensitive ecosystem?", *Progress in Oceanography*, vol. 44, no. 1, pp. 387–400, 1999.

[TÜR 08] TÜRKMEN M., TÜRKMEN A., TEPE Y. *et al.*, "Determination of metal contaminations in sea foods from Marmara, Aegean and Mediterranean seas: Twelve fish species", *Food Chemistry*, vol. 108, no. 2, pp. 794–800, 2008.

[ULU 19] ULUTURHAN E., DARILMAZ E., KONTAS A. *et al.*, "Seasonal variations of multibiomarker responses to metals and pesticides pollution in *M. galloprovincialis* and *T. decussatus* from Homa Lagoon, Eastern Aegean Sea", *Marine Pollution Bulletin*, vol. 141, pp. 176–186, 2019.

[URI 04] URIARTE I., VILLATE F., "Effects of pollution on zooplankton abundance and distribution in two estuaries of the Basque coast (Bay of Biscay)", *Marine Pollution Bulletin*, vol. 49, no. 3, pp. 220–228, 2004.

[VAL 06] VALAVANIDIS A., VLAHOGIANNI T., DASSENAKIS M. *et al.*, "Molecular biomarkers of oxidative stress in aquatic organisms in relation to toxic environmental pollutants", *Ecotoxicology and Environmental Safety*, vol. 64, pp. 178–189, 2006.

[VAN 15] VANDERMEERSCH G., LOURENÇO H.M., ALVAREZ-MUÑOZ D. *et al.*, "Environmental contaminants of emerging concern in seafood – European database on contaminant levels", *Environmental Research*, vol. 143, pp. 29–45, 2015.

[VAN 17] VAN DEN BERG H., MANUWEERA G., KONRADSEN F., "Global trends in the production and use of DDT for control of malaria and other vector-borne diseases", *Malaria Journal*, vol. 16, no. 1, pp. 401–401, 2017.

[VEN 02] VENTURA E.C., GAELZER L.R., ZANETTE J. *et al.*, "Biochemical indicators of contaminant exposure in spotted pigfish (*Orthopristis ruber*) caught at three bays of Rio de Janeiro coast", *Marine Environmental Research*, vol. 54, no. 3, pp. 775–779, 2002.

[VIA 91] VIARENGO A., CANESI L., PERTICA M. *et al.*, "Age-related lipid peroxidation in the digestive gland of mussels: The role of the antioxidant defence systems", *Experientia*, vol. 47, no. 5, pp. 454–457, 1991.

[VID 02] VIDAL M.-L., BASSÈRES A., NARBONNE J.-F., "Seasonal variations of pollution biomarkers in two populations of *Corbicula fluminea* (Müller)", *Comparative Biochemistry and Physiology Part C: Toxicology & Pharmacology*, vol. 131, no. 2, pp. 133–151, 2002.

[VIL 20] VILLARINO E., IRIGOIEN X., VILLATE F. *et al.*, "Response of copepod communities to ocean warming in three time-series across the North Atlantic and Mediterranean Sea", *Marine Ecology Progress Series*, vol. 636, pp. 47–61, 2020.

[VIÑ 18] VIÑAS L., PÉREZ-FERNÁNDEZ B., SORIANO J.A. *et al.*, "Limpet (*Patella* sp) as a biomonitor for organic pollutants. A proxy for mussel?", *Marine Pollution Bulletin*, vol. 133, pp. 271–280, 2018.

[WAN 16] WANG W.X., "Bioaccumulation and biomonitoring" in BLASCO J., CHAPMAN P.M., CAMPANA O., HAMPEL M. (eds), *Marine Ecotoxicology*, Academic Press, Elsevier, Amsterdam, 2016.

[WEY 10] WEYDERT C.J., CULLEN J.J., "Measurement of superoxide dismutase, catalase and glutathione peroxidase in cultured cells and tissue", *Nature Protocols*, vol. 5, no. 1, pp. 51–66, 2010.

[YIG 18] YIGIT M., CELIKKOL B., YILMAZ S. *et al.*, "Bioaccumulation of trace metals in Mediterranean mussels (*Mytilus galloprovincialis*) from a fish farm with copper-alloy mesh pens and potential risk assessment", *Human and Ecological Risk Assessment: An International Journal*, vol. 24, no. 2, pp. 465–481, 2018.

[YUA 20] YUAN Y., SUN T., WANG H. *et al.*, "Bioaccumulation and health risk assessment of heavy metals to bivalve species in Daya Bay (South China Sea): Consumption advisory", *Marine Pollution Bulletin*, vol. 150, p. 110717, 2020.

[ZHE 18] ZHELYAZKOV G., YANKOVSKA-STEFANOVA T., MINEVA E. *et al.*, "Risk assessment of some heavy metals in mussels (*Mytilus galloprovincialis*) and veined rapa whelks (*Rapana venosa*) for human health", *Marine Pollution Bulletin*, vol. 128, pp. 197–201, 2018.

7

Possible Interactions Between Environmental Microplastics, Copepods (*Tigriopus brevicornis*) and Oysters (*Crassostrea gigas*) in the Pertuis Charentais Sea

In this study, we examine and observe the presence of invasive copepods in Pertuis Charentais seawater (Pertuis Breton, France) under controlled laboratory conditions. During an experiment carried out on adult oysters (*Crassostrea gigas*, [BAY 19, BAY 17]), reared in an aquarium with seawater from Pertuis Charentais and exposed to environmental microplastics, we were able to observe a large number of easily identifiable populations of copepods: *Tigriopus brevicornis* [MÜL 76]. Several questions then arise regarding the observation of this species, which was referenced a few years ago around shellfish concessions and in the interval liquids of oysters. Potentially, this copepod could play a competitive role in the predation and filtration of phytoplankton by oysters. Some effects have been demonstrated with other copepod species on oyster gonads, but also on the filtration capacity in blue mussels (*Mytilus edulis*, [JON 09]). In addition, in view of our experiments on microplastics (MPs), could the copepods be a link in the trophic chain, potentially capable of transporting agglomerates of MPs in the respiratory and digestive system of the oyster? This work could pave the way for observations

---

Chapter written by Arno BRINGER, Emmanuel DUBILLOT, Grégoire PRUNIER and Hélène THOMAS.
For a color version of all figures in this book, see www.iste.co.uk/muttin/marine.zip.

on the presence and identification of this species in the Pertuis Charentais region, as well as on the development of oysters.

7.1. Introduction

France is the second-largest shellfish farmer in Europe. Each year, it produces an average of 200,000 tons of crustaceans, generating a turnover of roughly 774 million euros [CNC 20]. Shellfish farming is a key element of the Charente-Maritime coast. Whether it is farmed mussels in Pertuis Breton or oysters in the Marennes-Oléron region – growing and maturing in offshore breeding farms and clearings – the work of the farmers has shaped the landscape and generated major economic activity, making the region famous [GIL 14].

Each year, nearly 130,000 tons of oysters are produced and sold in France. The Charente-Maritime shellfish sector represents the largest French production area (with 22% of national production). It generates around 25,000 direct jobs for an annual turnover of roughly 300 million euros (most of which comes from the Marennes-Oléron region [MIL 11]). Nonetheless, shellfish farming has been encountering difficulties for several decades. Recently, the situation has worsened. Since 2008, massive mortality events in oysters have been observed. In 2014, 2016 and 2018, the mussel sector in Poitou-Charentes experienced high-mortality episodes, affecting 80–90% of production. Oyster and mussel farmers were left wondering about their annual production losses, which was documented even before the 21st century [GOU 98]. Concurrently, in the past decades, the constant increase in plastics production has led to an accumulation of plastic waste in aquatic ecosystems, particularly in coastal areas. Since their persistence and toxicity were proven, the study of plastic components has become of public interest [COL 11, CAP 18, BEI 18]. The degradation of plastic waste can produce microplastics (MPs < 5 mm) and then nanoplastics (NPs $< 1\mu m$, [NOA 08]). Nevertheless, scientific studies have only recently focused on the toxicity of NPs [CHA 17]. Although their mechanical and physiological impacts have been highlighted [COL 15], MPs are emerging solid substrates in marine environments. They are able to absorb chemical contaminants, but also to potentially interact with the trophic chain, transporting and vectorizing these chemical components on living organisms.

In this context, our research focused on the interactions between the environment and the physiology of bivalves. These are sentinel species for the study of pollution and environmental interactions [RIN 10]. The objective is to link the oyster mortality events observed by professionals to potential pollution from MPs. Furthermore, the presence of marine copepods in shellfish farming areas could be a sign of potential interaction in their trophic chain. Indeed, a previous study indicated that bivalves were able to feed on mesoplanktonic organisms, such as marine copepods [PRI 05, pp. 31–51]. In summer, trophic competition between copepods and oysters could impact the oysters' survival in the Pertuis Charentais area. Until this point, no one had studied the Pertuis Charentais Sea and the potential trophic relationships between copepods, oysters (*Crassostrea gigas*) and MPs.

Under controlled laboratory conditions, we studied and observed the presence of invasive copepods in Pertuis Charentais seawater. The experiment involved adult oysters (*Crassostrea gigas*, [BAY 19, BAY 17]) reared in tanks filled with seawater from Pertuis Charentais, and exposed to environmental MPs. We identified a large number of copepods: *Tigriopus brevicornis* [MÜL 76]. We considered the observations on them, previously documented around shellfish concessions and in the liquid interval of oysters. The copepods could play a competitive role in the predation and filtration of phytoplankton by oysters. As yet, this phenomenon has not been documented for *Crassostrea gigas*. The impact of other copepod species on oyster gonads and blue mussel filtration capacity (*Mytilus edulis*, [JON 09]) has already been highlighted. In addition, according to our experiments, copepods, which are links in the trophic chain, were able to transport MP agglomerates to the respiratory and digestive systems of oysters. This work could open the path to observing and identifying species in the Pertuis Charentais Sea, as well as studying the oysters' development. Another study indicated the key role of the food chain in transporting MPs [SET 14].

7.2. Materials and methods

7.2.1. *Study area*

Our site was located in the CNRS experimental marsh (LIENSs, La Rochelle, France). This marsh had a regular supply of seawater from Pertuis Breton (Figure 7.1) which was therefore used in our experiments. Shellfish (mussels and oysters, [BRO 94]) farming is an important economic activity in the Pertuis Breton area. The area is mainly supplied by water from the Sèvre Niortaise and Lay rivers (France). The Pertuis Breton area includes the coast of La Rochelle, one of the biggest French mussel production sites, and the island of Île de Ré, known for its oyster cultures. In the Pertuis Charentais, new structural adjustments in mussel production zones have enabled culture growth [RIC 00]. Freshwater inflows are relatively small, with maximum flows in winter barely exceeding 100 m^3 per second [DAR 04].

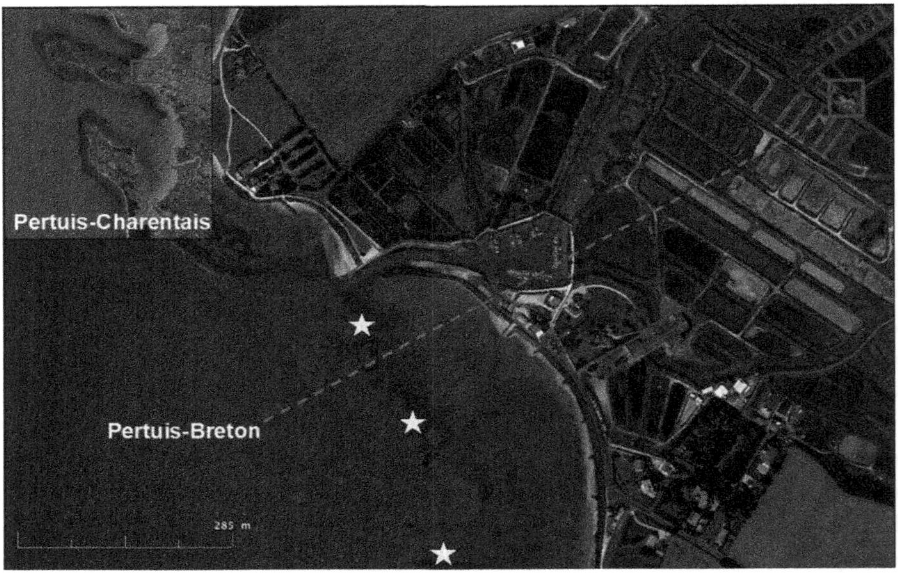

Figure 7.1. *Map of the Pertuis Charentais with a close-up (red square) of Pertuis Breton (the seawater area) and of the experimental site of oyster breeding (blue square). The pink arrow represents the seawater alimentation for the oysters' experimental development. The yellow stars show the locations of oyster farming*

7.2.2. *Experimental design*

Under controlled laboratory conditions and after a seven-day acclimatization period, adult oysters (24-month old, 15.75 ± 1.5 g, France Naissain hatchery) were reared in a tank (37.5 L, three replicates by condition and 30 oysters/tank) for two months (June to August 2019). This experiment was carried out on the CNRS experimental marsh premises (LIENSs, La Rochelle). Exposure to environmental microplastics (MPs) – polyethylene

(PE), polypropylene (PP) and PVC, 138.6 ± 2.3 μm – was set up with the following three conditions: control (without MPs); 0.1 mg MP. L^{-1} and 10 mg MP. L^{-1}. The MP mix consisted of 28% PE, 40% PP and 32% PVC [AND 11] corresponding to the MP levels found in the environment.

Regular water renewal was carried out using 10-μm filtered seawater (Figure 7.2) from the Pertuis Charentais (Figure 7.1, Pertuis Breton, France). The Pertuis Breton seawater was received directly on site in order to create similar environmental conditions to those of the shellfish professionals. In addition, seawater is rich in phytoplankton and bacteria, guaranteeing a prolific environment for the well-being and proper development of oysters. A thermoregulator containing UV (flow rate 78.3 $L.h^{-1}$) and constant bubbling were installed to homogenize the basins and eliminate potential bacteria. The physico-chemical parameters were constantly monitored. The oysters were kept at a temperature of 17.5 ± 0.5°C and salinity of 35 ± 0.5. The pH levels, nitrates and chlorine were kept constant and monitored for the duration of the experiment. Oysters were fed once a day with a mixed diet (*T. Iso, Pavlova, Tetraselmis, Thalassiosira* and *Nanno*, Shellfish Diet®). A pumping system was set up in each buffer tank to homogenize MPs in the water.

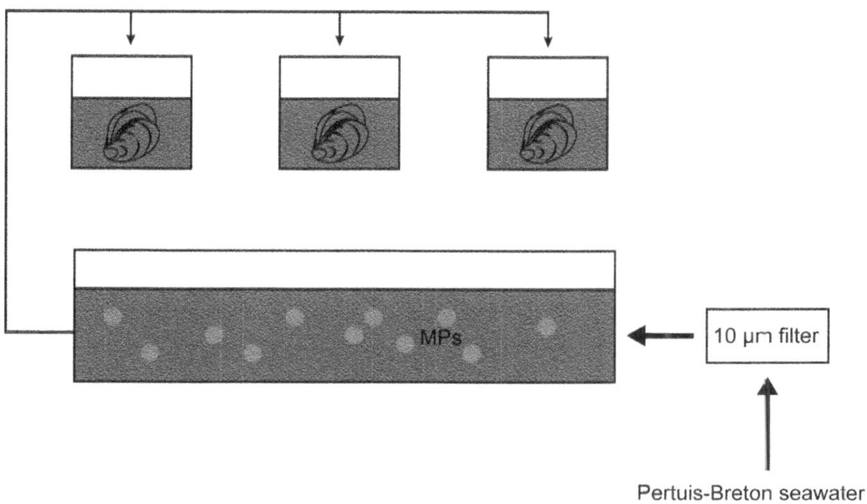

Figure 7.2. *Experimental design of a large tank filled with MPs supplying three tanks containing oysters. The large tank is renewed with 10-μm filtered seawater from Pertuis Breton*

7.2.3. *Sampling and observation methods*

Every two weeks (frequency of observations), samples were collected in our experimental seawater and in the interval liquid of oysters. After recovering the samples, which were doped with formaldehyde (Sigma-Aldrich, France) at 1%, the recovered volumes were kept cool (4°C). We added 25 μL of formalin in the 2-mL sampled volumes. This enabled us to fix the organisms in the samples. Subsequently, observations were made using the ZEISS Axio Observer Z1 microscope (x20) and pictures could be taken.

7.3. Results and discussions on the presence of copepods in seawater and in oysters

After a few weeks of experimentation, oyster mortality events were noted (high mortality with 10 mg MP. L^{-1} conditions). The MP concentrations tested were higher than the concentrations found in the environment [BEI 18]. Nonetheless, they were similar to sites with high concentrations of plastics. The goal was to better understand the MP concentrations in the Pertuis Charentais area. At the end of the experiment, the mortality rates were 2.2% for the controlled conditions, 13.3% for oysters exposed to 0.1 mg MP. L^{-1} and 23.3% for oysters exposed to 10 mg MP. L^{-1}.

An oyster was therefore opened to observe and understand the phenomenon more fully. High concentrations of microorganisms were highlighted. Inter-valve fluid samplings were collected to identify and analyze persistent parasites in dead oysters. Bibliographic research enabled us to identify the copepods (Figures 7.3 and 7.4) found in seawater, on the oyster gills and in the inter-valve fluid. These were *Tigriopus brevicornis* [MÜL 76], belonging to the family Harpacticidae Dana, 1846. Copepods are mesozooplankton, which bivalves ingest in large quantities during filtration [DAV 00]. A high concentration of *T. brevicornis* was found in oysters (Figures 7.3 and 7.4), up to 100 individuals/mL. Both male (Figure 7.3) and female (Figure 7.4) species had equal representation. Naupliar developmental N1 stages (Figure 7.4, [RAI 07]) were noticed. Indeed, females had ovigerous sacs (Figure 7.4).

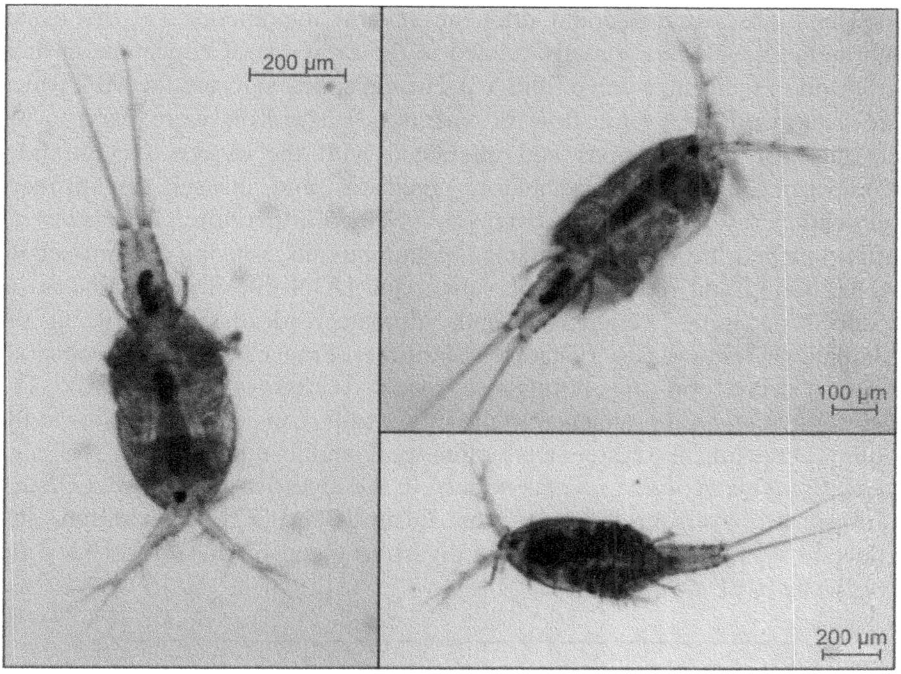

Figure 7.3. *Observations and pictures (ZEISS Axio Observer Z1, x20) of male copepods in seawater*

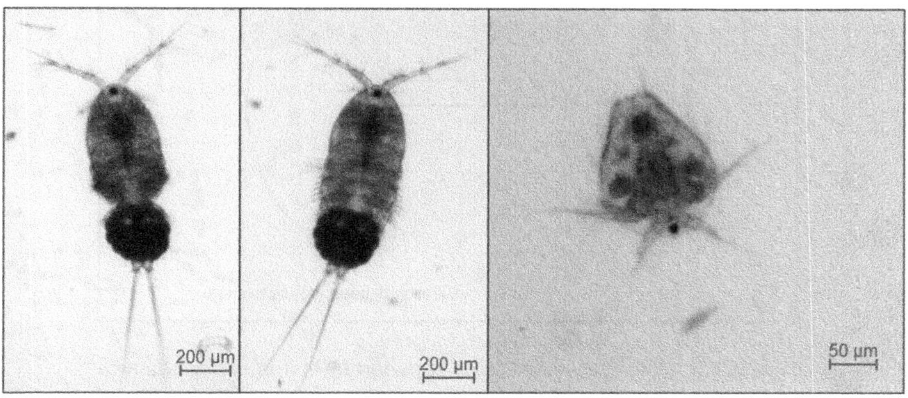

Figure 7.4. *Observations and pictures (ZEISS Axio Observer Z1, x20) of female copepods and larva in seawater*

These observations could draw on several hypotheses: (1) the oyster mortality events were directly related to the presence of copepods on their gills and in their inter-valve fluid. (2) The copepods ingested the MPs which are vectorized by trophic flow in oysters. (3) The MPs were fixed to the antennae of the copepods and interacted with the oysters by filtration. Competition between zooplankton organisms and mussels for primary production resources were proven [DAV 00]. Furthermore, *T. brevicornis* often touched the inhalant tentacles of the mussels, causing the mussels to stop filtering and reducing their valve gape [JON 09]. This could lead to reduced nutrition, weight loss and, ultimately, death. In 1970, in the Marennes-Oléron basin (Charente-Maritime, France), parasitic copepods were observed on the Portuguese oyster (*Crassostrea angulata*). The *Pseudomyicolidae sp.* attached to the oyster gills, causing perforations at the point of attachment. Oysters were often weak and lean [COM 72]. The same year, *Pseudomyicolidae sp.* was noticed in the digestive gland of flat oysters (*Ostrea edulis*) from the Arcachon Basin [COM 72]. In addition, this copepod was also identified in the digestive gland of the mussel (*Mytilus edulis*) by [LED 64].

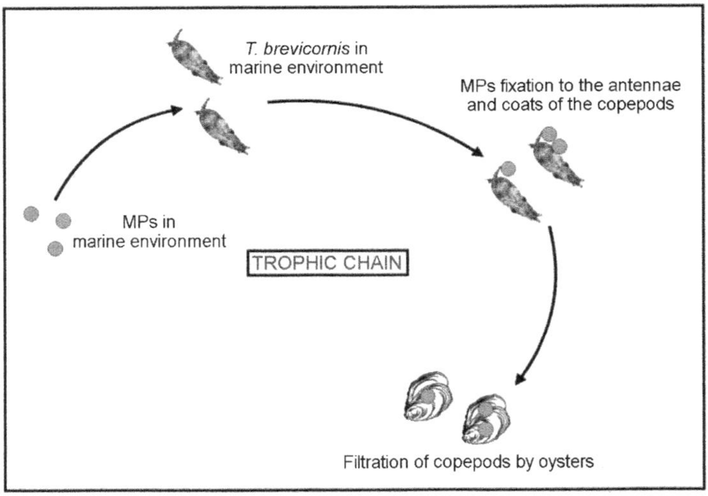

Figure 7.5. *Potential interactions between MPs and copepods with possible fixation of MPs to the copepods. Then, through the food chain, the oysters filter nutritional resources in seawater*

Nevertheless, under controlled conditions (without MPs), mortality events were not recorded. Thus, MPs could be attached to the antennae of *Tigriopus brevicornis* and directly ingested by oysters (Figure 7.5). It was previously shown that MPs become fixed to the tentacles of Hydrozoire, *H. attenuata* [MUR 18]. Several papers suggested that MPs, such as micro- and nano-sized beads, may have negative impacts on marine copepods [LEE 13]. Other studies on *Tigriopus japonicus* and the ingestion of fluorescent beads underlined that copepods entered the food chain directly and readily ingested MPs [LEE 13, COL 13].

7.4. General conclusion and perspectives

Exposure of oysters to environmental MPs enabled us to observe a high concentration of *T. brevicornis* copepods in oysters (*C. gigas*). Mortality events are not directly linked to copepods. Nonetheless, it is important to discover whether or not copepods are key hosts of MPs, which could be transmitted to oysters when filtering the mesozooplankton. In addition, bivalves have the ability to select filtered particles. Oysters could therefore recognize MPs and release them through pseudo-feces, without having digested them [WAR 19]. However, if MPs are inside the copepod stomach, the oysters cannot detect the MPs and therefore digest both the copepod and the MPs. Additional studies will need to be developed to understand these processes and the toxic action potential of MPs for filter feeders. These observations were presented to shellfish professionals. They are interested in pursuing this work and in understanding the copepods' role in the oyster food chain in the Pertuis Charentais area.

7.5. References

[AND 11] ANDRADY A.L., "Microplastics in the marine environment", *Marine Pollution Bulletin*, vol. 62, no. 8, pp. 1596–1605, 2011.

[BAY 17] BAYNE B.L., AHRENS M., ALLEN S.K. *et al.*, "The proposed dropping of the genus *Crassostrea* for all Pacific cupped oysters and its replacement by a new genus *Magallana*: A dissenting view", *Journal of Shellfish Research*, vol. 36, no. 3, pp. 545–547, 2017.

[BAY 19] BAYNE B., D'AURIAC M.A., BACKELJAU T. *et al.*, "A scientific name for Pacific oysters", *Aquaculture*, vol. 499, p. 373, 2019.

[BEI 18] BEIRAS R., BELLAS J., CACHOT J. *et al.*, "Ingestion and contact with polyethylene microplastics does not cause acute toxicity on marine zooplankton", *Journal of Hazardous Materials*, vol. 360, pp. 452–460, 2018.

[BRO 94] BROSSARD S., Détermination des sels nutritifs limitant la production primaire dans les bassins conchylicoles du Pertuis Breton et de Marennes-Oléron, DUT Thesis, IFREMER La Tremblade and La Rochelle University Institute of Technology, La Rochelle, 1994.

[CAP 18] CAPOLUPO M., FRANZELLITTI S., VALBONESI P. *et al.*, "Uptake and transcriptional effects of polystyrene microplastics in larval stages of the Mediterranean mussel *Mytilus galloprovincialis*", *Environmental Pollution*, vol. 241, pp. 1038–1047, 2018.

[CHA 17] CHAE Y., AN Y.J., "Effects of micro- and nanoplastics on aquatic ecosystems: Current research trends and perspectives", *Marine Pollution Bulletin*, vol. 124, no. 2, pp. 624–632, 2017.

[CNC 20] CNC, Comité National de la Conchyliculture, available at: http://www.cnc-france.com/La-Production-francaise.aspx [accessed April 2020], 2020.

[COL 11] COLE M., LINDEQUE P., HALSBAND C. *et al.*, "Microplastics as contaminants in the marine environment: A review", *Marine Pollution Bulletin*, vol. 62, no. 12, pp. 2588–2597, 2011.

[COL 14] COLE M., WEBB H., LINDEQUE P.K. *et al.*, "Isolation of microplastics in biota-rich seawater samples and marine organisms", *Scientific Reports*, vol. 4, p. 4528, 2014.

[COL 15] COLE M., GALLOWAY T.S., "Ingestion of nanoplastics and microplastics by Pacific oyster larvae", *Environmental Science & Technology*, vol. 49, no. 24, pp. 14625–14632, 2015.

[COM 72] COMPS M., "Sur un copépode parasite de l'huître portugaise (*Crassostrea angulata* Lmk) dans le bassin de Marennes-Oléron", *Congrès International pour l'Exploration de la Mer, Comité des Crustacés, Coquillages et Benthos,* vol. K, no. 35, 1972.

[DAR 04] DARDIGNAC-CORBEIL M.J., "La mytiliculture dans le Pertuis Breton : synthèse des travaux réalisés de 1980 à 1992", *Annales de la Société des sciences naturelles de la Charente-Maritime*, vol. 10, pp. 3–79, October 2004.

[DAV 00] DAVENPORT J., SMITH R.J., PACKER M., "Mussels *Mytilus edulis*: Significant consumers and destroyers of mesozooplankton", *Marine Ecology Progress Series*, vol. 198, pp. 131–137, 2000.

[GIL 14] GILBERT R., DURET J., "Recensement de la conchyliculture 2012. La Charente-Maritime est le premier département producteur de coquillages", *Agreste Poitou-Charentes*, vol. October, no. 16, 2014.

[GOU 98] GOULLETQUER P., SOLETCHNIK P., LE MOINE O. *et al.*, "Summer mortality of the Pacific cupped oyster *Crassostrea gigas* in the Bay of Marennes-Oléron (France)", *ICES Mariculture Committee*, pp. 14–21, Copenhagen, Denmark, June, 8–13 1998.

[JON 09] JONSSON A., NIELSEN T.G., HRUBENJA I., "Eating your competitor: Functional triangle between turbulence, copepod escape behavior and predation from mussels", *Marine Ecology Progress Series*, vol. 376, pp. 143–151, 2009.

[LED 64] LE DANTEC J., BOREL M., "Les moules du littoral atlantique au sud de la Gironde et leurs crustacés parasites (observations d'octobre 1963 à avril 1964 sur la rive gauche de la Gironde et dans le bassin d'Arcachon)", *Conseil International pour l'Exploration de la Mer, Comité des Mollusquès et Crustacés*, no. 34, 1964.

[LEE 13] LEE K.-W., SHIM W.J., KWON O.Y. *et al.*, "Size-dependent effects of micro polystyrene particles in the marine copepod *Tigriopus japonicus*", *Environmental Science & Technology*, vol. 47, no. 19, pp. 11278–11283, 2013.

[MIL 11] MILLE D., LE MOINE O., Adaptabilité des activités conchylicoles aux modifications de leur environnement : scénarii et solutions. Le cas du bassin de Marennes-Oléron, Project, Projet ANCORIM, CREAA–IFREMER, 2011.

[MÜL 76] MÜLLER O.F., *Zoologiæ Danicæ Prodromus, seu Animalium Daniæ et Norvegiæ indigenarum characteres, nomina, et synonyma imprimis popularium*, vol. 1, nos i–viii, pp. 1–170, Hallageri, Copenhagen, 1776.

[MUR 18] MURPHY F., QUINN B., "The effects of microplastic on freshwater *Hydra attenuata* feeding, morphology & reproduction", *Environmental Pollution*, vol. 234, pp. 487–494, 2018.

[NOA 08] NOAA, *Proceedings of the International Research Workshop on the Occurrence, Effects, and Fate of Microplastic Marine Debris*, University of Washington Tacoma, Tacoma, USA, September 9–11, 2008.

[PRI 05] PRINS T., ESCARAVAGE V., "Can bivalve suspension-feeders affect pelagic food web structure?", in DAME R.F., OLENIN S. (eds), *The Comparative Roles of Suspension-Feeders in Ecosystems*, Springer, Dordrecht, 2005.

[RAI 07] RAISUDDIN S., KWOK K.W., LEUNG K.M. *et al.*, "The copepod *Tigriopus*: A promising marine model organism for ecotoxicology and environmental genomics", *Aquatic Toxicology*, vol. 83, no. 3, pp. 161–173, 2007.

[RIC 00] RICHARD P., Etude des potentialités conchylicoles du Pertuis Breton : mise en évidence de l'utilisation des différentes sources de matière organique par traçage isotopique naturel, Final report, Contrat no. 98000-127, Direction Départementale de l'Agriculture et de la Forêt de Charente-Maritime, 2000.

[RIN 10] RINGWOOD A.H., MCCARTHY M., BATES T.C. *et al.*, "The effects of silver nanoparticles on oyster embryos", *Marine Environmental Research*, vol. 69, pp. S49–S51, 2010.

[SET 14] SETÄLÄ O., FLEMING-LEHTINEN V., LEHTINIEMI M., "Ingestion and transfer of microplastics in the planktonic food web", *Environmental Pollution*, vol. 185, pp. 77–83, 2014.

[WAR 19] WARD J.E., ROSA M., SHUMWAY S.E., "Capture, ingestion, and egestion of microplastics by suspension-feeding bivalves: A 40-year history", *Anthropocene Coasts*, vol. 2, no. 1, pp. 39–49, 2019.

8

Health Assessment of Juvenile Scallops in the La Rochelle Harbor

8.1. Introduction

For a long time, fears have been raised about heavy metal contamination, mainly in harbor areas. Human activities generate waste, which is dissolved in particles, macro-waste and liquid effluent. Previous studies have highlighted that scallop species (and the assessment of their digestive glands) were great indicators of the environmental quality of harbors. To limit contamination and preserve the marine environment, regulations were put in place (ISO 9 001, ISO 14 001, OHSAS 18 001).

Moreover, it was noted that *M. varia* species could be used as biomarkers in biomonitoring models, to assess environmental pollution. Research on adult *M. varia* revealed a rapid bioaccumulation of heavy metals and a similar toxicity to previous studies carried out *in situ* in the La Rochelle marina [BRE 20]. Thus, the study developed low-cost and convertible methods to monitor the water quality in different areas, such as the La Rochelle harbor. Juvenile scallops were used as an indicator species to assess the diffusion of pollution in harbor environments (careening and rainwater areas).

As illustrated in Figure 8.1, juvenile marine bivalve species were biomonitored using an outdoor caging technique. This work aims to observe

---

Chapter written by Marine Paomia BREITWIESER, Marine BARBARIN, Angélique FONTANAUD and Hélène THOMAS.

For a color version of all figures in this book, see www.iste.co.uk/muttin/marine.zip.

the impact of inorganic chemical contamination on the health status of juvenile marine bivalve *Mimachlamys varia*.

Since many papers have been dedicated to the adult black scallop *M. varia*, this experiment on juvenile species was highly innovative. Heavy metal contaminants were assessed and the following health biomarkers were measured: Laccase (immune reaction), GST (detoxification process) and MDA (lipid peroxidation).

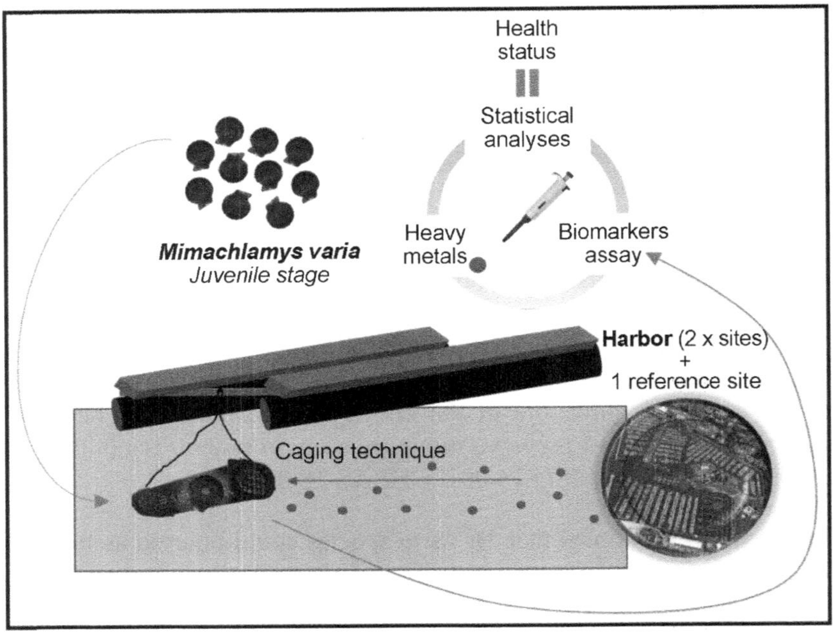

Figure 8.1. *Integrative approach to assess the health status of juvenile scallops in the La Rochelle harbor*

8.2. Assessment methods of the health status of marine bivalves

8.2.1. *Sampling sites and caging transplantation procedure*

Three sampling sites were chosen to assess the health status of juvenile marine bivalves:

– the reference site, which was a closed marsh of the laboratory, located a few kilometers away from the La Rochelle marina;

– the harbor careening site, which was exposed to contamination due to boat handling (anti-fouling, paint residues, sacrificial anodes), although it possessed a filtration system;

– the harbor rainwater site, which reflected pollution from rainwater (run-off from roads, etc.).

The same sites were used for previous studies on the adult black scallop *Mimachlamys varia* [BRE 20]. In 2016, some six-month-old *M. varia* scallops from the Tinduff hatchery (Brittany region) were placed in traps at our reference site for a fortnightly acclimatization. Once the traps were on-site, sampling was carried out regularly for the biochemical analysis and quantification of heavy metals. On day 0 (D0), 30 scallops were sampled from the reference site. On day 7 (D7) and day 21 (D21), 30 scallops were sampled from all of the sites. Individuals were stored at -80°C, as detailed in the research paper [BRE 20].

8.2.2. *Water physicochemical assessments*

During the sampling period and on the sampling dates, the temperature and salinity were measured at all of the study sites, using a multi-parameter probe (YSI 6600V2), in line with the method used in a previous study [BRE 20].

8.2.3. *Trace element assessments*

Trace elements analyses were carried out according to previous studies [MIL 15, BRE 17]. The analyses of 14 trace elements (Ag, As, Cd, Co, Cr, Cu, Fe, Mn, Ni, Pb, Se, Sn, V, Zn) were performed using two measuring instruments: a Varian Vista-Pro ICP-OES and a Thermofisher Scientific XSeries 2 ICP-MS.

8.2.4. *Biomarker assessments*

Biomarker assays (MDA, GST and Laccase) were performed on the SAFAS Flx - Xenius. On each sampling date (D0, D07, D21), tests were carried out on pools (n = 3) of juvenile scallops (whole body). The results for

GST and Laccase were expressed in the UI/mf protein and the results for MDA were expressed in the μM/mg protein. The total protein concentrations were assessed using the BCA Kit methodology (Bicinchononique Acid Kit, Sigma Aldrich). The BCA Kit contained bovine serum albumin (BSA) as standard, and the reduction of Cu2 Cuby alkaline proteins [SMI 85] at absorbance 562 nm.

8.2.4.1. *Glutathione S-transferase (GST) assay*

The glutathione S-transferase (GST) specific activity plays a key role in xenobiotic detoxification. The GST activity (GST being a group of enzymes) was determined using the Sigma method kit (CS0410-1 KT) [BRE 16].

8.2.4.2. *Malondialdehyde (MDA) assay*

Oxidative stress was assessed through lipid peroxidation by quantifying malondialdehyde (MDA), a chemical metabolite of lipid degradation. The MDA concentration was measured using a commercially available MDA assay kit (Oxis International) [MIL 15].

8.2.4.3. *Laccase assay*

The phenoloxidase (PO) activity (Laccase) revealed immune system alterations in bivalves. Its enzymatic activity is contaminant sensitive [LUN 10] and plays a crucial role in the immune defense mechanism of marine invertebrates.

8.2.5. *Statistical analysis*

The statistical analysis was performed (see Figure 8.2) using R software (R Core Team, 3.2.2 version) [RCO 20]. The survey depended on the sampling dates (three dates), the chosen sites (three sites) and the size of the samples for each sub-group (10 pools of three specimens).

For all variables, the homogeneity of variance was assessed using Fligner-Killeen tests. Then, at the three sampling sites and for the three sampling dates (D0, D7, D21), the multivariate response of individuals was measured using a Multivariate Factorial Analysis.

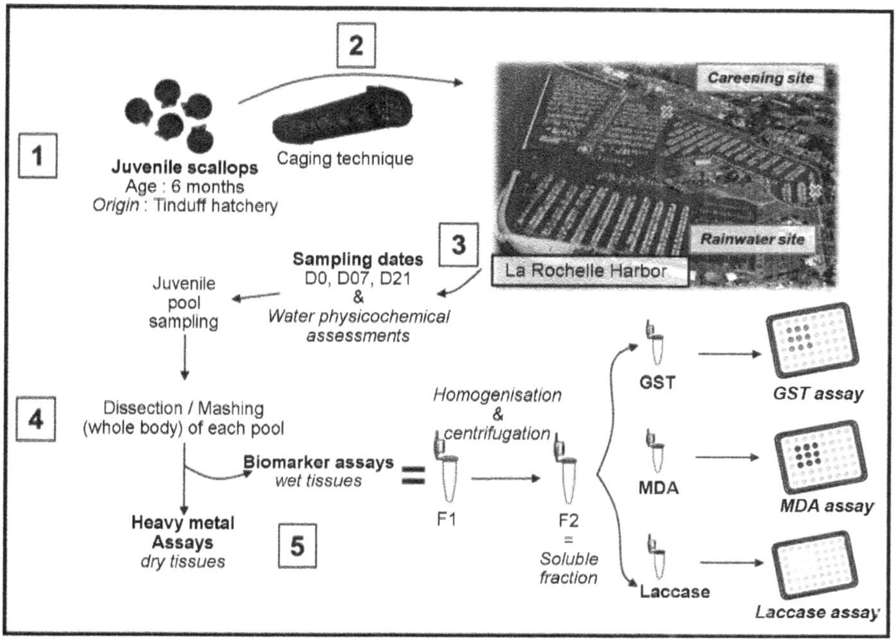

Figure 8.2. *Illustration of the monitoring station, including the sampling procedure and the biomarker and heavy metal assays*

8.3. Results and discussion on the health status of juvenile scallops

The following section merges both the results and discussion of our defined parameters, in order to assess the health status of juvenile scallops and the quality of the three study sites.

8.3.1. *Water physical variables*

The water physical variables are presented in Table 8.1. The temperature trends ranged from 16.8°C to 19.7°C, which represented the weather conditions. In contrast, the salinity depended on the climatic conditions and fluctuated from 27.61 to 33.19 PSU. This phenomenon, explained by rainfall bringing fresh water and sediments, was mainly observed at the rainwater station on day 21.

	Marsh = Reference site			Careening			Rainwater		
Day	D0	D7	D21	D0	D7	D21	D0	D7	D21
Temp (°C)	16.80	17.90	19.70	17	16.82	19.01	16.80	16.95	18.92
Sal (PSU)	32.22	33.12	33.19	31.96	31.94	32	30.54	31.61	27.61

Table 8.1. *Physicochemical parameters at the three sampling sites: marsh (reference site), careening and rainwater areas. Temperature (Temp) is measured in degree Celsius and salinity (Sal) is measured in Practical Salinity Unit (PSU)*

8.3.2. *Health biomarker assessments*

In Table 8.2 and Figure 8.3, the enzymatic and concentration assays testified the health status of juvenile marine bivalves. These analyses were carried out to assess the detoxification capacities of xenobiotics (GST) and determine the integrity of cell membranes, the involvement of lipid peroxidation (MDA), and the immune system modulations (Laccase).

Laccase activity, GST activity and MDA content are common biomarkers in ecotoxicology and biomonitoring studies [VIA 07, IUM 13, NAH 13].

In terms of the Laccase modulation, which explains immune system alterations, the results were as follows:

– When comparing the three sites on day 21, there was:

- no difference between the marsh and careening area;

- a significant difference ($p < 0.05$) between the reference site (marsh, 4.28 ± 0.99 UI/mg of proteins) and rainwater area (3.60 ± 0.57 UI/mg of proteins);

- a significant difference ($p < 0.05$) between the careening (4.61 ± 0.95 UI/mg of proteins) and rainwater areas (3.60 ± 0.57 UI/mg of proteins).

– When comparing the three sampling dates (between day 0 and day 21) there was:

- a significant increasing signal of 34% ($p < 0.05$) for the reference site (marsh);

- a significant increasing difference ($p < 0.05$) in the immune system (45%) for the harbor careening site;

- a significant increasing difference ($p < 0.05$) in the immune system (13%) for the harbor rainwater site.

	Marsh = Reference site				Careening				Rainwater			
	Day 0	Day 07	**Day 21**	*D0-D21*	Day 0	Day 07	**Day 21**	*D0-D21*	Day 0	Day 07	**Day 21**	*D0-D21*
Laccase *immune system alteration*	3.18 ± 0.82 UI/mg of proteins	4.36 ± 1.15 UI/mg of proteins	4.28 ± 0.99 UI/mg of proteins *[a]	↑ 34 %*	3.18 ± 0.82 UI/mg of proteins	4.36 ± 1.15 UI/mg of proteins	4.61 ± 0.95 UI/mg of proteins *[a]	↑ 45 %*	3.18 ± 0.82 UI/mg of proteins	4.26 ± 0.72 UI/mg of proteins	3.60 ± 0.57 UI/mg of proteins *[b]	↑ 13 %* (↓15% D7-D21)
GST *xenobiotic detoxification*	0.79 ± 0.2 UI/mg of proteins	1.08 ± 0.14 UI/mg of proteins	1.39 ± 0.32 UI/mg of proteins *[a]	↑ 76 %*	0.79 ± 0.22 UI/mg of proteins	0.78 ± 0.23 UI/mg of proteins	0.76 ± 0.25 UI/mg of proteins [b]	n.d	0.79 ± 0.22 UI/mg of proteins	1.12 ± 0;31 UI/mg of proteins	0.78 ± 0.26 UI/mg of proteins [b]	n.d
MDA *lipid peroxidation*	38.06 ± 4 µM/mg of proteins	75.3 ± 19 µM/mg of proteins	69 ± 22 µM/mg of proteins *[a]	↑ 82 %*	38.06 ± 4 µM/mg of proteins	33.80 ± 6 µM/mg of proteins	24.35 ± 4 µM/mg of proteins *[b]	↓ 37 %*	38.06 ± 4 µM/mg of proteins	38.90 ± 13 µM/mg of proteins	40.79 ± 7 µM/mg of proteins [c]	n.d

Table 8.2. *Biomarker results in juvenile scallops at each study site on day 0, day 7 and day 21. For each date, significant differences between the impacted sites and the marsh (reference site) are indicated using asterisks: *$p < 0.05$. The values represent the mean ± standard error (n = 7 pools of three individuals per site). For each site and biomarker, D0-D21 is the difference between day 0 and day 21. For each site on day 21, values that have no significant difference ($p > 0.05$) share the same letter. No letters are shown when there was no significant statistical difference among sites*

In the marsh and careening areas, the non-specific immune system (defined as the last line of defense after the physicochemical barriers, mucus, byssus and cutaneous epithelia) functioned up to day 21. On the last sampling day (day 21), an 18% drop in the immune system was observed at the rainwater site, in comparison with the other two sites.

During the early stages of the study, the results suggested that there was no oxidative stress or reactive oxygen species neutralized by a mixture of contaminants (e.g. pesticides, hydrocarbons, heavy metals), mainly coming from waste of watershed and urban activities [BRE 20, LUS 11].

In terms of GST activity, which assesses the detoxification capacities, the results were as follows:

– When comparing the three sites on day 21, there was:

- a significant decreasing difference ($p < 0.05$) between the marsh (1.39 ± 0.32 UI/mg of proteins) and careening area (0.76 ± 0.25 UI/mg of proteins);

- a significant decreasing difference ($p < 0.05$) between the marsh (1.39 ± 0.32 UI/mg of proteins) and rainwater area (0.78 ± 0.26 UI/mg of proteins);

- no difference between the careening (0.76 ± 0.25 UI/mg of proteins) and rainwater areas (0.78 ± 0.26 UI/mg of proteins).

– When comparing the three sampling dates (between day 0 and day 21) there was:

- a significant increasing signal of 76% ($p < 0.05$) for the reference site (marsh);

- no significant difference between the three sampling dates for the harbor careening site;

- no significant difference between the three sampling dates for the harbor rainwater site.

A higher specific activity was observed at the reference site (marsh) in comparison with the other two sites. In addition, the enzymatic activity increased over time (+ 76% between day 0 and day 21). Over time, there was no variation at the harbor careening and rainwater sites.

In comparison with literature [BRE 20], the results for juvenile scallops contradicted the results for adult scallops. Thus, one may ask if GST activity is a relevant biomarker in juvenile scallops or species at early stages. At a premature age, it is likely that scallops may not possess a Phase II type of conjugation activity, or an excretion activity by extra-cellular mobilization.

In terms of MDA content, which determines the integrity of cell membranes and the involvement of lipid peroxidation, the results were as follows:

– When comparing the three sites on day 21, there was:

- a significant difference ($p < 0.05$) between the reference site (marsh, 69 ± 22 μM/mg of proteins) and careening area (24.35 ± 4 μM/mg of proteins);

- a significant difference ($p < 0.05$) between the reference site (marsh, 69 ± 22 μM/mg of proteins) and rainwater area (40.79 ± 7 μM/mg of proteins);

- a significant difference ($p < 0.05$) between the careening (24.35 ± 4 μM/mg of proteins) and rainwater area (40.79 ± 7 μM/mg of proteins).

– When comparing the three sampling dates (between day 0 and day 21) for each site, there was:

- a significant increasing MDA concentration of 82% ($p < 0.05$) for the reference site (marsh);

- a significant decreasing difference ($p < 0.05$) in the immune system (37%) for the harbor careening site;

- no significant difference between day 0 and day 21 for the harbor rainwater site.

This result highlighted decreased lipid peroxidation in scallops from the two harbor sites. A study on oysters from the reference site noted similar results for juvenile mollusks [SÉG 17]. Before starting the experiment, the scallops were acclimatized for a week. On day 21, the low level of MDA content at both the careening and rainwaters areas, when compared to the reference site, could suggest that the acclimatization period was not long enough. At the end of the experiment, the increased MDA content levels at the reference site could imply a better metabolization of black scallops and thus, poor detection at the time [SÉG 17].

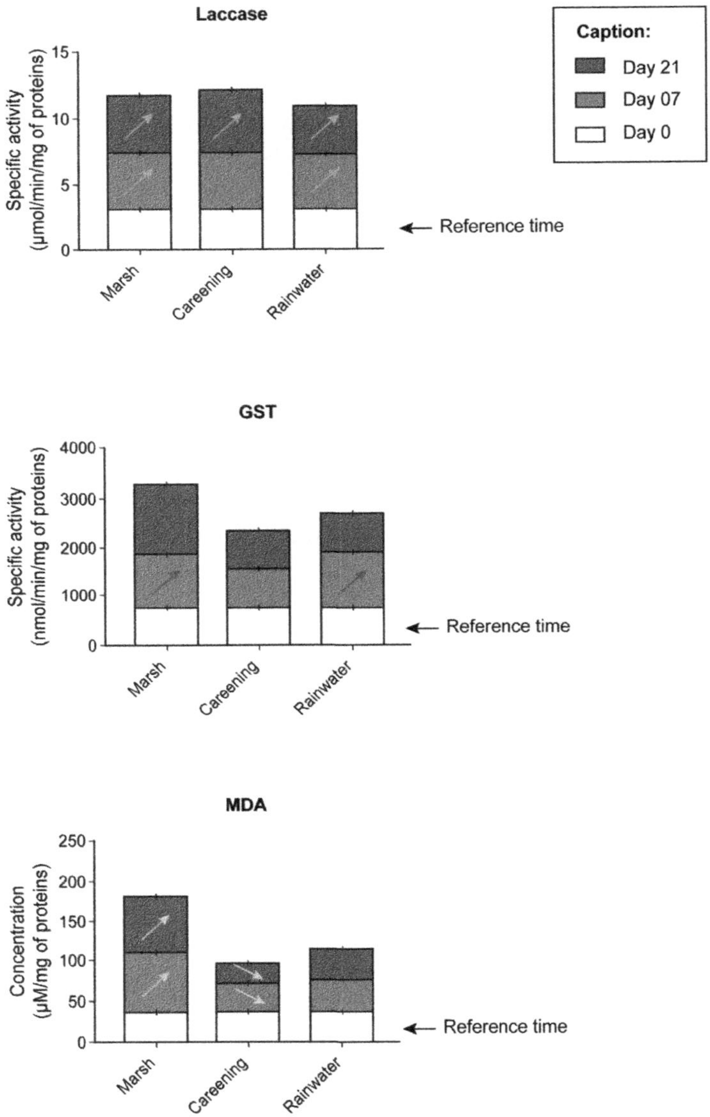

Figure 8.3. *Histograms illustrating the biomarker results in juvenile* M. varia *at the reference site (marsh) and both harbor sites (rainwater and careening areas) on D0, D07 and D21 (n = 7 pools per site and date). For each site and biomarker, the arrows represent the difference in the biomarker level between D0 and D7 or D21. The absence of arrows means that, for each site, there is no difference between the days, as shown by the GST activity and MDA content at the careening and rainwater sites, respectively*

The MDA content, evaluated in the juvenile stage of *M. varia*, indicates that the biomarker responding to the accumulation of peroxidation products reveals a membrane destabilization linked to the decrease in the capacity of oxidative defenses [SÉG 17].

To conclude this part, all of the biomarker results contradicted previous studies on adult scallops from the same sites. Furthermore, as described in literature, the temperatures and abiotic factors were optimum for the good development of juvenile scallops [PEC 07]. Therefore, the hypothesis that abiotic factors were confounding factors was excluded. In addition, the juvenile scallops came from a hatchery (outside of the three study sites) with a known history of life, so life adaptation to each environment was also excluded. This suggested that juvenile scallops did not benefit from adequate biomarker responses. Thus, they could not be used as markers to assess the quality of the marine environment. For early stage species, alternatives could be considered, such as the study of genetic diversity. Stressors may cause genotoxicity, which is sensitive to alterations in the DNA structure and functional genomics/gene functions. As a result, the cellular and tissue malfunctions could cause physiological disturbances, having repercussions on the population. It could impact both reproduction and development, which has been proven in previous studies on similar species [BRE 18].

The standardization of the "Comet Assay" [SMI 06] could help when assessing the genotoxic impacts of chemical molecules on neoplastic factors (random mutations, etc.).

8.3.3. *Heavy metals*

Between D0 and D21, the heavy metal assays showed clear variations in concentrations, depending on the experimental sites (careening, rainwater and reference sites) (Table 8.5).

When looking at the reference site (marsh) on day 21, there was:

– a significant decreasing difference ($p < 0.05$) in the Cr concentration (56%) for the marsh site only;

– a significant increasing difference ($p < 0.05$) in concentrations of Co (53%), Pb (21%), Se (38%) and Mn (400%).

At the three sites, the Cr concentrations obtained were lower than those previously recorded in the adult black scallop *M. varia* [BRE 20]. The Cr concentrations decreased at the reference site on D21 and lower values could be associated with the Tinduff hatchery, where the juvenile scallops were from.

The Co and Se results were lower than those obtained in previous studies on adult stage scallops. On the other hand, the Pb results were higher (Table 8.3):

	Marsh	**Careening**	**Rainwater**
Juvenile (6 months)	2.96 ± 0.08	2.5 ± 0.13	2.42 ± 0.24
Adult	2.1 ± 0.2	1.7 ± 0.3	1.4 ± 0.2
Difference Juvenile/Adult (%)	41	47	73

Table 8.3. *Comparison of the Pb results at juvenile and adult stages (data from [BRE 20])*

Another study had already identified specific Pb pollution at this site [BRE 18]. A greater sensitivity to Pb absorption in juvenile scallops could explain the differences with the adult species. There was a significant increase in Mn concentrations at the careening and rainwater sites (62% for both). On D21, when comparing the juvenile and adult results [BRE 20], there was a clear difference in Mn concentration levels (Table 8.4).

	Marsh	**Careening**	**Rainwater**
Juvenile (6 months)	211.03 ± 24.25	68.16 ± 4.57	68.41 ± 11.37
Adult	40.26 ± 6.1	28.32 ± 4.6	28 ± 4.5
Difference Juvenile/Adult (%)	424	141	144

Table 8.4. *Comparison of the Mn results at juvenile and adult stages (data from [BRE 20])*

The results for adult scallops were similar to the results from previous papers studying the same species [BRE 17]. Mn is a key element in the synthesis of several enzymes, such as Mn-SOD. This trace element, which might be present in anti-fouling paints, could be toxic in high concentrations. The results suggested that, for certain trace elements including Mn and Pb, the bioaccumulation capacity of juvenile scallops was greater than adult scallops. Thus, the juvenile black scallop did not bioaccumulate heavy metals in a similar manner to the adult species.

When looking at the careening and rainwater areas on day 21, there was:

– a significant increasing difference *($p < 0.05$)* in Cu (careening: 3,560%, rainwater: 2,555%), Fe (careening: 57%, rainwater: 90%), Zn (careening: 55%, rainwater: 41%) and Mn (Table 8.4);

– a significant decreasing difference *($p < 0.05$)* in As (careening: 13%, rainwater: 20%), Cd (careening: 20%) and Se (rainwater: 21%).

The Cu results were similar for both the juvenile and adult stages, even if the concentrations remained lower [BRE 20]. The Cu increase could be explained by a specific Cu pollution. This was already identified in a previous study at the La Rochelle marina [BRE 18]. The Zn results were higher in juvenile scallops than in adult scallops. This could suggest that juvenile scallops have a greater sensitivity to Zn, Mn and Pb. The Fe results were lower for juvenile scallops at the careening site and higher in the rainwater area [BRE 20]. Zn, Cu and Fe are notably present in anti-fouling paints, harbor structures and cathodic protection with sacrificial anodes. Thus, they are part of the harbor environment. The differences in Fe concentrations between the two stages of black scallop development and between the careening and rainwater sites could be explained by the heavy rains that took place during the experiment. For many years, La Rochelle and its marina have been implementing measures to protect the environment, including installing filtration devices in harbor basins. Some are located on the careening site to treat wastewater containing anti-fouling paint residues, and others are located on the rainwater outlet of the harbor. Although the efficiency of the devices has been proven, heavy rains increase run-off from roads (cars, plastics, etc.). This phenomenon could explain the results in juvenile scallops.

On D21, the results remained lower at the juvenile stage than at the adult stage. The Cd results were much lower in juvenile scallops than in adult scallops [BRE 20]. Thus, at the juvenile stage, scallops may be less sensitive to these trace elements. A decrease in Se (although not significant) was observed at the careening site. A significant increase was noted at the reference site (Table 8.5). Although the Se levels in juvenile scallops were lower than in adult scallops, the trend at the reference site was similar, with higher Se concentrations [BRE 20]. Similar to the levels of Pb, this suggested a specific presence at the reference site (marsh).

	Marsh = Reference site				Careening				Rainwater			
	Day 0	Day 07	**Day 21**	*D0–D21*	Day 0	Day 07	**Day 21**	*D0–D21*	Day 0	Day 07	**Day 21**	*D0–D21*
As	16.81 ± 0.64	17.84 ± 0.67	16.92 ± 0.42	ns	16.81 ± 0.64	15.19 ± 0.70	14.62 ± 0.32 [a,1]	↓13%	16.81 ± 0.64	14.68 ± 0.81	13.41 ± 0.18 [a,1,2]	↓20%
Cd	5.02 ± 0.26	5.45 ± 0.43	5.44 ± 0.40	ns	5.02 ± 0.26	4.47 ± 0.23	4.03 ± 0.17 [a,1]	↓20%	5.02 ± 0.26	4.95 ± 0.37	3.93 ± 0.48 [1]	ns
Co	0.49 ± 0.03	0.57 ± 0.03	0.75 ± 0.03 [a]	↑53%	0.49 ± 0.03	0.57 ± 0.03	0.59 ± 0.03 [1]	ns	0.49 ± 0.03	0.65 ± 0.04	0.55 ± 0.04 [1]	ns
Cr	1.56 ± 0.24	0.61 ± 0.04	0.69 ± 0.08 [a]	↓56%	1.56 ± 0.24	1.54 ± 0.12	1.21 ± 0.08 [1]	ns	1.56 ± 0.24	1.35 ± 0.10	1.06 ± 0.12 [1]	ns
Cu	6.87 ± 0.24	8.17 ± 0.30	18.29 ± 9.41	ns	6.87 ± 0.24	88.47 ± 6.11	251.46 ± 18.58 [a,1]	↑3560%	6.87 ± 0.24	98.98 ± 10.89	182.46 ± 8.02 [a,1,2]	↑2555%
Fe	392.19 ± 20.94	370.46 ± 31.06	464.82 ± 53.78	ns	392.19 ± 20.94	581.21 ± 21.30	614.54 ± 36.88 [a]	↑57%	392.19 ± 20.94	754.25 ± 67.01	745.71 ± 71.38 [a,1]	↑90%
Mn	42.11 ± 3.35	62.21 ± 5.28	211.03 ± 24.25 [a]	↑400%	42.11 ± 3.35	52.29 ± 2.77	68.16 ± 4.57 [a,1]	↑62%	42.11 ± 3.35	60.29 ± 3.22	68.41 ± 11.37 [a,1]	↑62%
Ni	1.60 ± 0.13	1.37 ± 0.06	1.64 ± 0.06	ns	1.60 ± 0.13	1.54 ± 0.09	1.30 ± 0.05 [1]	ns	1.60 ± 0.13	2.32 ± 0.78	1.35 ± 0.10 [1]	ns
Pb	2.44 ± 0.15	2.56 ± 0.17	2.96 ± 0.08 [a]	↑21%	2.44 ± 0.15	2.24 ± 0.07	2.50 ± 0.13	ns	2.44 ± 0.15	2.71 ± 0.15	2.42 ± 0.24	ns
Se	6.22 ± 0.27	10.09 ± 0.50	8.58 ±0.28 [a]	↑38%	6.22 ± 0.27	5.87 ± 0.30	5.36 ± 0.19 [1]	ns	6.22 ± 0.27	5.94 ± 0.32	4.93 ± 0.12 [a,1]	↓21%
V	NA	NA	NA	ns	NA	NA	1.43 ± 0.00	ns	NA	NA	NA	ns
Zn	559.60 ± 58.39	480.61 ± 62.72	595.50 ± 52.28	ns	559.60 ± 58.39	687.11 ± 55.49	869.03 ± 65.91 [a,1]	↑55%	559.60 ± 58.39	697.84 ± 70.15	791.21 ± 54.19 [a]	↑41%
Ag	1.15 ± 0.16	1.32 ± 0.24	0.94 ± 0.10	nd	1.15 ± 0.16	0.79 ± 0.07	0.80 ± 0.06	nd	1.15 ± 0.16	1.28 ± 0.17	0.87 ± 0.07	nd
Sn	16.81 ± 0.64	17.84 ± 0.67	16.92 ± 0.42	ns	16.81 ± 0.64	15.19 ± 0.70	14.62 ± 0.32 [a,1]	↓13%	16.81 ± 0.64	14.68 ± 0.81	13.41 ± 0.18 [a,1,2]	↓20%

Table 8.5. *Heavy metal results in juvenile scallops (µg/g of dry weight) at each study site on day 0, day 7 and day 21. On each sampling date, significant differences for the two impacted sites and the marsh (reference site) are indicated by: a ($p < 0.05$). On D21, the significant differences for the three sites are indicated by 1 (marsh/careening and marsh/rainwater), 2 (careening/rainwater) ($p < 0.05$). The values represent the mean ± standard error (n = 7 pools of three individuals per site). For each site and heavy metal, D0–D21 is the difference between day 0 and day 21. When comparing each site on day 21, the values that have no significant difference ($p > 0.05$) share the same letters (ns)*

When comparing the juvenile and adult stages of the black scallop *M. varia*, the results indicated variations in the response to heavy metals. Depending on the sites, a greater sensitivity to Pb, Mn, Zn or Fe was observed at the juvenile stage.

8.3.4. *Principal components analysis*

A descriptive and multi-criteria analysis of the data obtained on juvenile bivalves is presented in Figure 8.4. The analysis focused on the first two components, as they were responsible for 48.42% of the variance (Axis 1: 27.78%, Axis 2: 20.64%), which was sufficient to perform the analysis.

In terms of biomarkers, GST activity and MDA content were positively correlated, unlike Laccase which reacted independently. Regarding metals, a strong correlation between Sn, Cu, Zn and Fe was observed. At the two harbor sites on D21, the last three elements showed a significant increase in concentration. They were negatively correlated with As, Cd, Se and Ag. At the two harbor sites on D21, the first three elements showed a significant decrease in concentration. On the other hand, the concentration of Mn, which also showed a significant increase at the three sites on D21, was not correlated with Sn, Cu, Zn and Fe. Indeed, it was correlated with Co and Pb, which showed a significant increase in concentration at the reference site on D21. The GST activity was positively correlated with Pb, Co and Mn, and Laccase with Co. The observations in Figure 8.4A corresponded to those in Figure 8.4B, where the reference site (marsh) stood out from the two harbor sites.

The multifactorial analysis divided metals and biomarkers into four groups. Amongst them, group two stood out, having a strong correlation between Co, Pb, Mn and GST. At the reference site (marsh) on D21, the four elements showed a significant increase in concentration/activity. Figure 8.4B pictures correlation circles for each site. The marsh stood out from the two harbor sites (careening and rainwater). Thus, the detoxification process of juvenile bivalves, although not properly established, could be involved. Group four showed a strong correlation between Sn, Cu, Zn, Fe and Laccase. At the careening and rainwater sites, these elements were similar to previous results. Finally, groups two and four, and groups one and three were negatively correlated.

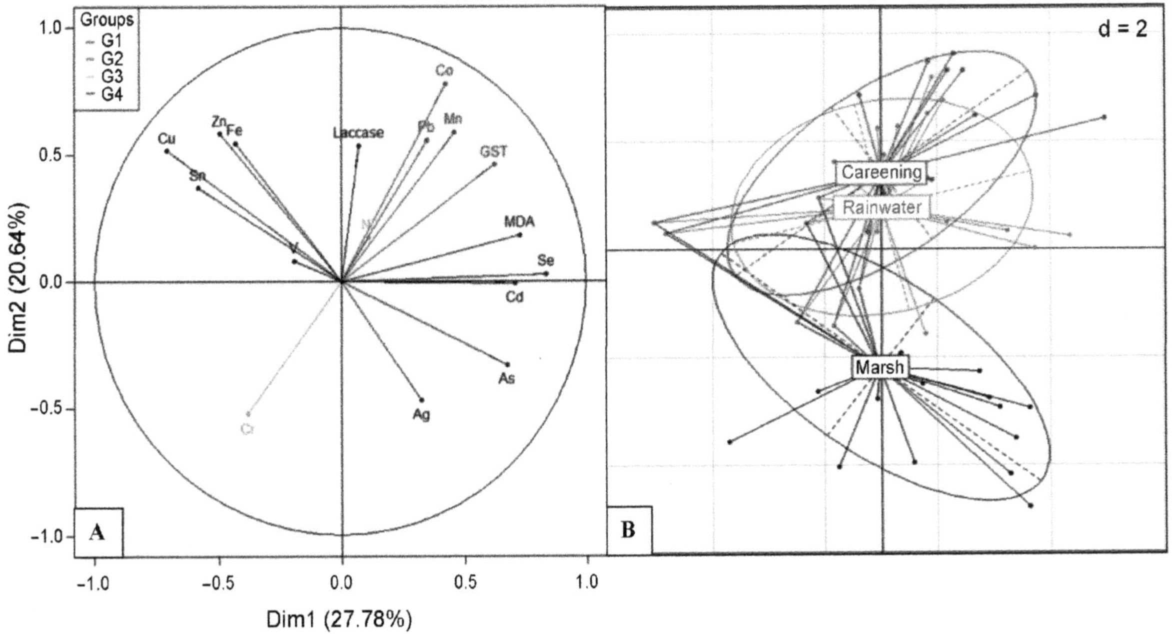

Figure 8.4. *Multifactorial analysis graph highlighting individual responses to biomarkers and heavy metals at the three sampling sites (marsh, careening and rainwater) for 21 days. The first two axes indicate a variance of 48.42% (Axis 1: 27.78%, Axis 2: 20.64%). MDA: MDA content, GST: GST activity, Laccase: Laccase activity*

MDA was positively correlated with GST. On the other hand, Laccase was not correlated with the other two biomarkers. The results showed that the detoxification process, as well as the lipid peroxidation, were only active at the marsh site (although a significant decrease in MDA concentration was noticed at the careening site). On the other hand, Laccase showed immune activity, which was not optimum, at all sites.

There was a clear difference between the reference site and the careening and rainwater experimental sites. The detoxification process and the immune system of the black scallop *M. varia* at the juvenile stage (six months) were not necessarily optimum. This work, which was a preliminary study, monitored juvenile scallops for the first time. At this stage of development, black scallops may be more fragile and sensitive to their environment. Although, given the differences between the study sites, they can adapt. Another study on the hollow oyster *Crassostrea gigas* at various stages of development, highlighted an increase in phenoloxidase activity when aging [THO 09], which proved a growth in immune activity with age.

Other studies on juvenile species are necessary to expand our knowledge of six-month bivalves. Our results confirmed that adult black scallops *M. varia* were well suited for biomonitoring studies, particularly in harbor areas facing cathodic protection, and in assessing the potential effects of sacrificial anodes on the health of bivalves.

8.4. References

[BRE 16] BREITWIESER M., VIRICEL A., GRABER M. *et al.*, "Short-term and long-term biological effects of chronic chemical contamination on natural populations of a marine bivalve", *PLoS ONE*, vol. 11, no. 3, p. e0150184, 2016.

[BRE 17] BREITWIESER M., VIRICEL A., CHURLAUD C. *et al.*, "First data on three bivalve species exposed to an intra-harbor polymetallic contamination (La Rochelle, France)", *Comparative Biochemistry and Physiology Part C: Toxicology & Pharmacology*, vol. 199, pp. 28–37, 2017.

[BRE 18] BREITWIESER M., DUBILLOT E., BARBARIN M. *et al.*, "Assessment of the biological quality of port areas: A case study on the three harbors of La Rochelle: The marina, the fishing harbor and the seaport", *PLoS ONE*, vol. 13, no. 6, p. e0198255, 2018.

[BRE 20] BREITWIESER M., BARBARIN M., PLUMEJEAUD-PERREAU C. *et al.*, "Biomonitoring of *Mimachlamys varia* transplanted to areas impacted by human activities (La Rochelle Marine, France)", *Chemosphere*, vol. 243, p. 125199, 2020.

[IUM 13] IUMMATO M.M., DI FIORI E., SABATINI S.E. *et al.*, "Evaluation of biochemical markers in the golden mussel *Limnoperna fortunei* exposed to glyphosate acid in outdoor microcosms", *Ecotoxicology and Environmental Safety*, vol. 95, pp. 123–129, 2013.

[LUN 10] LUNA-ACOSTA A., ROSENFELD E., AMARI M. *et al.*, "First evidence of laccase activity in the Pacific oyster *Crassostrea gigas*", *Fish & Shellfish Immunology*, vol. 28, no. 4, pp. 719–726, 2010.

[LUS 11] LUSHCHAK, V.I., "Environmentally induced oxidative stress in aquatic animals", *Aquatic Toxicology*, vol. 101, no. 1, pp. 13–30, 2011.

[MIL 15] MILINKOVITCH T., BUSTAMANTE P., HUET V. *et al.*, "*In situ* evaluation of oxidative stress and immunological parameters as ecotoxicological biomarkers in a novel sentinel species (*Mimachlamys varia*)", *Aquatic Toxicology*, vol. 161, pp. 170–175, 2015.

[NAH 13] NAHRGANG J., BROOKS S.J., EVENSET A. *et al.*, "Seasonal variation in biomarkers in blue mussel (*Mytilus edulis*), Icelandic scallop (*Chlamys islandica*) and Atlantic cod (*Gadus morhua*): Implications for environmental monitoring in the Barents Sea", *Aquatic Toxicology*, vol. 127, pp. 21–35, 2013.

[PEC 07] PECK L.S., POWELL D.K., TYLER P.A., "Very slow development in two Antarctic bivalve mollusks, the infaunal clam *Laternula elliptica* and the scallop *Adamussium colbecki*", *Marine Biology*, vol. 150, no. 6, pp. 1191–1197, 2007.

[RCO 20] R CORE TEAM, R: A language and environment for statistical computing, R Foundation for Statistical Computing, Vienna, Austria, available at: https://www.r-project.org/, 2020.

[SÉG 17] SÉGUIN A., MOTTIER A., PERRON C. *et al.*, "Sub-lethal effects of a glyphosate-based commercial formulation and adjuvants on juvenile oysters (*Crassostrea gigas*) exposed for 35 days", *Marine Pollution Bulletin*, vol. 117, nos 1–2, pp. 348–358, 2017.

[SMI 85] SMITH P.K., KROHN R.I., HERMANSON G.T. *et al.*, "Measurement of protein using bicinchoninic acid", *Analytical Biochemistry*, vol. 150, no. 1, pp. 76–85, 1985.

[SMI 06] SMITH C.C., O'DONOVAN M.R., MARTIN E.A., "hOGG1 recognizes oxidative damage using the comet assay with greater specificity than FPG or ENDOIII", *Mutagenesis*, vol. 21, no. 3, pp. 185–190, 2006.

[THO 09] THOMAS-GUYON H., GAGNAIRE B., BADO-NILLES A. *et al.*, "Detection of phenoloxidase activity in early stages of the Pacific oyster *Crassostrea gigas* (Thunberg)", *Developmental and Comparative Immunology*, vol. 33, no. 5, pp. 653–659, 2009.

[VIA 07] VIARENGO A., DONDERO F., PAMPANIN D.M. *et al.*, "A biomonitoring study assessing the residual biological effects of pollution caused by the HAVEN wreck on marine organisms in the Ligurian Sea (Italy)", *Archives of Environmental Contamination and Toxicology*, vol. 53, no. 4, pp. 607–616, 2007.

List of Authors

Marine BARBARIN
LIENSs CNRS
La Rochelle University
France

François BÉNITEZ
LIENSs CNRS
La Rochelle University
France

Marine Paomia BREITWIESER
LIENSs CNRS
La Rochelle University
France

Arno BRINGER
LIENSs CNRS
La Rochelle University
France

Pascal BRUNELLO
LIENSs CNRS
La Rochelle University
France

Rose CAMPBELL
EIGSI
La Rochelle
France

Carine CHURLAUD
LIENSs CNRS
La Rochelle University
France

Emmanuel DUBILLOT
LIENSs CNRS
La Rochelle University
France

Angélique FONTANAUD
Marina Harbor
La Rochelle
France

Lovina FULLGRABE
Station de recherches
sous-marines et océanographiques
de Calvi (STARESO)
Calvi
France

Sylvie GOBERT
Station de recherches
sous-marines et océanographiques
de Calvi (STARESO)
Liège University
Belgium

Marianne GRABER
LIENSs CNRS
La Rochelle University
France

Valérie HUET
LIENSs CNRS
La Rochelle University
France

Jean-Rémy HUGUET
LIENSs CNRS
La Rochelle University
France

Stéphane LE FLOCH
CEDRE
Brest
France

Michèle LEDUC
Station de recherches
sous-marines et océanographiques
de Calvi (STARESO)
Calvi
France

Pierre LEJEUNE
Station de recherches
sous-marines et océanographiques
de Calvi (STARESO)
Calvi
France

Adélaïde LEREBOURS
LIENSs CNRS
La Rochelle University
France

Michel MARENGO
Station de recherches
sous-marines et océanographiques
de Calvi (STARESO)
Calvi
France

Alice MAZEAUD
LIENSs CNRS
La Rochelle University
France

Frédéric MUTTIN
EIGSI
La Rochelle
France

Marion PILLET
LIENSs CNRS
La Rochelle University
France

Grégoire PRUNIER
LIENSs CNRS
La Rochelle University
France

Christine PLUMEJEAUD
LIENSs CNRS
La Rochelle University
France

Yann RABUTEAU
ALLEGANS Network
Brest
France

Anaïs RIDEL
LIENSs CNRS
La Rochelle University
France

Benoit SIMON-BOUHET
LIENSs CNRS
La Rochelle University
France

Hélène THOMAS
LIENSs CNRS
La Rochelle University
France

Index

A

abiotic, 50
absorbance, 44, 45
acetylcholine esterase (AChE), 108, 118
Adour-Garonne, 25
advective, 79
agency, 25, 27, 29, 37, 38, 93–95
agriculture, 28, 29
alert system, 28
aliquots, 60
alkaline, 44
anchors, 6
Angoulins, 43, 46–48
ANOVA, 45
anthropic, 56
anthropogenic, 41, 47
anti-fouling, 56, 159, 168, 169
antioxidant, 107, 117, 118
apoptosis, 107
aquaculture, 22, 29
asbestos, 114
atmospheric pressure, 60

B

bacteria, 21, 36, 149
  Escherichia coli, 36
  intestinal enterococci, 36
barriers, 1, 2, 4–6, 8, 10, 11
  floating, 91, 98
basin, 23–26, 28, 31, 32
  committee, 25
  river, 24–26, 28, 32
  sub-, 23
bathymetry, 60
Bay of Biscay, 4, 12, 16
beach, 58, 79
beads
  fluorescent, 153
  micro- and nano-sized, 153
behavior, 89, 99–101
benthic, 42
bilge, 12, 56
bioaccumulation, 106, 109, 110, 112, 113, 116, 119, 122, 123
biocenose, 105

biodiversity, 21, 37, 38
bioindicators, 42, 105, 111
biomarkers, 41–46, 48–50, 57, 58, 60, 61, 63, 67–73, 77, 78, 84, 157–159, 161–163, 165–167, 171–173
  multi-, 84, 122, 124
  of effect, 106, 117
  of exposure, 106
biomonitoring, 41, 50, 157, 162, 173
biotic, 50
blooms, 122
Blue Brigade, 56
Blue Flag, 56
bollards, 11
boom, 1–16
  C-shape, 12
  V-shape, 9
bottom-up, 100
Brest, 42, 43
Brittany, 43, 47, 48
bunker, 8, 12
buoy, 3, 5, 8, 9, 11

C

cable, 5
caging, 157, 158
careening, 56, 58, 59, 67, 73, 79, 83, 157, 159, 162–173
  area, 43, 58
catalase (CAT), 107
catchment area, 23, 25
  river, 25
chain, 9, 12
Charente, 1, 4, 6, 12, 13, 16
  -Maritime, 1, 6, 12, 13, 16, 21, 22, 27, 29–31, 33, 37, 96, 97, 146, 152
Châtelaillon, 79
Chef de Baie, 6, 8, 13, 15, 16
chemical, 103, 104, 106, 114, 118
  cocktail, 105
  contaminants, 42, 50
  pollutants, 41
chlorine, 115
chromium (Cr), 65
ciliary motility, 106
cities, 89, 90, 96–100
classification, 28, 30, 34, 36
  declassification, 28, 36
conference, 10
conjugation, 165
contamination, 22, 28, 36, 39, 41–50
  chronic, 48
copepods, 145, 147, 150–153
  Harpacticidae, 150
  Pseudomyicolidae sp., 152
  Tigriopus brevicornis, 145, 147, 150, 153
  Tigriopus japonicus, 153
copper (Cu), 56, 65, 67, 77
correlation, 71, 72, 78
Corsica, 113, 114, 122, 124
crisis communication, 100
crops, 32
current, 2, 3, 5–7, 9–11, 14
currentology, 58, 75
curtain, 2, 5, 12
cytosolic, 77

D

decision aid system, 89, 91, 93, 98
decision support system, 5
deliverables, 90–93
Deparmental Directorate of Territories and the Sea (DDTM), 6
detoxification, 41, 44, 48, 49, 62, 67, 78, 107, 108, 114, 115, 158, 160, 162–164, 171, 173
  process, 158, 171, 173

dichlorodiphenyldichloroethylene (DDE), 116, 117, 123
dichlorodiphenyltrichloroethane (DDT), 116, 117
diffusion, 56, 60, 79, 80
  of pollution, 157
digestive glands, 41, 42, 44–50, 58, 62, 64, 66–70, 73, 74, 78
directive, 22–24, 27, 28, 30, 35, 48, 56, 63, 103, 119
discharges, 104, 108, 109, 115, 117, 123
DNA structure, 167
dogfish, 61
downscaling, 5
dredging, 55, 57–59, 65, 67, 69, 73, 74, 78
dry weight, 61, 63, 65, 76
dynamometer, 13

E

e-learning, 2
echo-sounders, 60
ecosystems, 41, 50
ecotoxicology, 84, 107, 122, 124
eddies, 79
edible, 28
education, 100
employee representatives, 26
Enable, Avoid, Shift, and Improve (EASI), 89, 98, 99
end-users, 6, 17
endocrine disruptors, 105
energetic metabolism, 107, 123
energy reserves, 107
English Channel, 16
Environmental Quality Standards (EQS), 103
enzymes, 44, 45, 47–49, 57, 61, 62, 69, 77, 78, 107, 118, 123
epipelagic, 109
epizootic disease, 30
equations
  Navier-Stokes, 4
  Saint-Venant, 4, 5
European Maritime and Fisheries Fund (EMFF), 124
experiment, 1, 2, 6–8, 10, 12–17

F

fairing area, 56–59, 65, 67, 69, 75, 78, 79
Falmouth, 1, 3, 8, 10, 16
farming, 21, 22, 27, 29, 30, 32–34, 38, 43
fatty acids, 49, 107, 118, 119
fence, 2, 5
fertilizers, 73
filtration, 145, 147, 150, 152
  filter organisms, 111
  filtering system, 58
fine fractions (FF), 44, 45
finite-difference, 96
finite-element, 5, 6, 91, 96
finite-volume, 6
Fisher, 60
fishing, 22, 28–30, 36
  fishermen, 33, 37
  overfishing, 41, 42
flood, 23, 98, 99
food
  chain, 105, 110, 119, 147, 152, 153
  security, 119
  web, 119
forecasting, 1, 4
free-surface, 5, 59
free radicals, 77
fueling station, 14, 57–59, 63, 65, 67, 69, 71, 78, 83
funding, 29, 37, 90

G

gametes, 112
gates, 12, 14
gene, 107
genetic, 42
geolocation, 2
gills, 41, 42, 44–50, 150, 152
Gironde, 4, 23, 37–39
glutathione, 44, 45, 47, 49
  disulfide (GSSG), 107
  peroxidase (GPx), 107
  reductase (GR), 107
  S-transferase (GST), 44, 46, 47, 49, 107, 118, 158–160, 162–166, 171–173
gonads, 112, 145, 147
governance, 21, 22, 25, 37
GPS, 8, 9, 12
gradient, 109, 114
grid, 4, 5, 11
  multigrid, 5
  unstructured, 4, 59, 60
guidelines, 41, 48, 50, 55, 90, 92

H

habitats, 41
harbor, 1, 3, 8–10, 12–16, 157–159, 162, 164–166, 169, 171, 173
harmonization, 35
hatcheries, 111, 112, 159, 167, 168
Haworthiopsis attenuata, 153
health, 42, 43, 50, 56, 71, 77, 103–107, 111, 117–123
  risk, 120, 121
heavy metals, 49, 57, 63, 64, 66, 68, 70, 71, 73–78, 157–159, 161, 164, 167, 169–172
herpes virus, 30
Holm's correction, 63
homeostasis, 77
homoscedasticity, 63
hydro-geographical, 23, 25
hydrodynamic, 1–4, 6, 7, 11, 12, 16, 58–60, 79, 82–84
hydrostatic, 4
  non-, 59

I

Île de Ré, 43, 58, 80, 148
immune
  reaction, 158
  system, 160, 162–165, 173
immunity, 49, 67, 78
immunomodulation, 46, 48, 49, 67, 78
indicators, 23, 34–36, 38
indices, 119, 120, 122
inertia, 71, 72
insecticides, 104, 117
integrated management, 22, 25
inter-valve fluid, 150, 152
interval liquid, 145, 150
intoxication, 77
intra-harbor, 56, 63, 65, 71, 73, 84
iron (Fe), 60, 61, 63, 65, 67, 71, 73
irrigation, 32, 33
ISO 14001 standard, 55, 56, 83
isodrine, 43

J, K

juvenile, 30, 157–159, 161–163, 165–171, 173
knowledge management, 93

L

La Rochelle, 1, 3–6, 8, 13–15, 55–60, 73, 75, 76, 79, 80, 82–84, 90, 93, 96, 97
Laccase, 45, 46, 48–50
Lagrangian, 7, 11
Lay River, 148
lead (Pb), 60, 61, 63, 65, 66, 71, 73, 76, 78

LIDAR, 60
lipid peroxidation, 62, 67, 69, 77, 107, 118, 158, 160, 162, 163, 165, 173
lipophilic, 114
Lisbon, 1, 3, 6–8, 10, 11, 16
load shackles, 8, 9
lobbyists, 32
local authorities, 25, 26, 29, 37
local consultation, 100
Loire-Bretagne, 25
lysosomal, 118

M

malondialdehyde (MDA), 45–49, 107, 158–160, 162, 163, 165–167, 171–173
manganese (Mn), 77
  superoxide dismutase (Mn-SOD), 168
Marennes-Oléron, 29, 146, 152
marina, 50, 115, 157, 158, 169
marine park, 37, 38
Marine Strategy Framework Directive, 103
maritime traffic, 115, 123
marker, 27
marsh, 147, 148, 158, 162–173
mayor, 27
media, 22, 30, 31, 33, 34
membranes, 5, 6, 107, 118
  cell, 77, 162, 165
  plasma, 77
mesh, 4–7, 16
mesoplanktonic, 147
mesozooplankton, 150, 153
metabolites, 57
metabolization, 165
metal-binding, 108
metallothionein (MT), 108
meteorology, 79, 80, 83, 84
metolachlor, 43
microorganisms, 150
microscope, 150
mid-ocean ridges, 104
mitochondria, 77
mobile application, 95
model
  business, 91, 93
  circulation, 60
  global tidal, 60
  hydrodynamic, 2, 4, 6, 11, 16, 59, 82, 83
modeling, 1, 3, 4, 9, 10
modulation, 42, 46, 49
mooring, 2, 3, 5, 6, 9–11, 14
mortality, 23, 30, 31, 33, 34, 36, 39, 146, 147, 150, 152, 153
mudflats, 78
multi-parameter probe, 159
multidisciplinary, 55
multifactorial, 171, 172
multivariate, 160
municipalities, 26, 29, 30
Mussel Watch, 106

N

Naupliar, 150
network, 56
neurotoxicity, 108, 118, 123
nickel (Ni), 60, 61, 63, 65, 71, 73, 76
normality, 63
North Atlantic Oscillation, 80
Nouvelle-Aquitaine, 22, 29, 42, 43, 48
numerical method, 5
nutritional interests, 33

O

ocean sciences, 101
OHSAS 18001 standard, 55
oil
  drifting, 16
  intermediate fuel, 8, 12

oligotrophic, 108, 109
Open-MI, 91
open-sea, 114, 117
open areas, 43, 50, 58
  semi-closed areas, 42, 58, 73, 83
  semi-open areas, 43, 75
organochlorine, 104, 115, 123
osmoregulatory, 107
OSPAR Convention, 42, 63
outermost regions, 24
outlet
  dredging, 43, 46, 47, 50, 55, 57–59, 65, 67, 69, 73, 74, 78
  fairing, 43, 46, 47
  rainwater, 43, 46, 47
outliers, 64–66, 68, 70
ovigerous sacs, 150
oxidant, 107
oxidation, 44, 45
  auto-, 45
oxidative stress, 48, 49
oxygen, 24, 41, 48, 49

P

Palmonés estuary, 96
para-chlorine, 115
participative science, 100
pathogens, 33
peroxidation, 45, 49
personal care products (PCPs), 104
Pertuis Breton, 80, 145–149
Pertuis Charentais, 21, 22, 25–27, 30, 31, 34–36, 39, 91, 96, 145–150, 153
  Sea, 91, 145, 147
Pertuis d'Antioche, 12
pesticides, 33, 34, 36, 78, 104, 117, 118
  organochlorine pesticides (OCPs), 104
petrogenic, 115
pH, 149
pharmaceuticals, 104
phenanthrene, 115
phenoloxidase (PO), 45, 49, 107, 118, 160, 173
photic layer, 108
phycotoxins, 28
physico-chemical, 24, 28, 35
phytoplankton, 28, 108, 109, 114, 122, 145, 147, 149
phytosanitary, 41
place of refuge, 14
plastic, 32
  microplastics (MPs), 34, 36, 145–149, 152, 153
  waste, 146
platform, 89, 94–96, 99, 100
Poitou-Charentes, 146
policies, 21–25, 27, 29–32, 38, 98–100
POLMAR, 96, 97
polybromodiphenyl (PBDE), 104, 116, 117
polychlorinated biphenyl (PCB), 104, 114–116, 118
polycyclic aromatic hydrocarbon (PAH), 75, 76, 78, 104, 114–116, 118, 123
polyethylene (PE), 149
polymetallic, 73
polypropylene (PP), 149
port
  fishing, 13
  of Caramiñal, 96
  river, 12
Prefect, 26
preparation, 1, 2, 4, 15, 17, 89, 94, 98, 99
prevention, 89, 99
primary production, 108, 109, 114, 152
principal component analysis (PCA), 63, 71, 72
priority substances, 103
protected area, 23, 37, 38

protein, 44, 45, 47, 48, 58, 61, 62, 67, 69, 160, 162–165
public, 21–23, 25–28, 30, 31, 33, 36–38
  health, 22, 27, 36, 38
PVC, 149

Q, R

quay, 3, 12, 14
reactive oxygen species (ROS), 49, 107, 110
real time, 9
regulations, 48, 49, 157
replicability, 90
Réseau d'Observation de la Contamination CHimique du littoral (ROCCH), 106
resilience, 100
resolution, 4–6, 11, 60
responders, 1, 6, 10
risk management, 100
Rochefort, 1, 6, 8, 12, 13, 16
run-off, 105, 115, 117

S

sacrificial anodes, 159, 169, 173
sailors, 56
salinity, 149, 159, 161, 162
sanitation, 29
sea–land, 37–39
seafood, 118, 119
seagrass, 105, 110
  Posidonia oceanica, 110
sediment, 48, 73, 76
sentinel species, 36, 42, 50, 106, 111, 147
separator, 83
serum, 44
Seudre, 4
Sèvre-Niortaise River, 4
sharing economy, 91
shellfish, 21–23, 27–39
  bivalves, 41, 42, 44, 48–50, 147, 150, 153
    burrowing, 28
  farming, 146–148
  gastropods, 28
  lobster, 61
  mussels, 30, 31, 33, 34, 42, 106, 111, 112, 114–119, 122, 124, 145–148, 152
    Mytilus edulis, 116, 145, 147, 152
    Mytilus galloprovincialis, 111–113, 116, 120, 121
  oysters, 23, 30, 33, 34, 36, 39, 42, 48, 145, 146–150, 152, 153
    Crassostrea gigas, 49, 145, 147, 153
    Ostrea edulis, 152
  scallops, 41, 42, 44, 45, 47–50, 56, 58, 59, 64, 66, 73, 74, 77, 78, 84, 157–159, 161, 163, 165, 167–171, 173
    Chlamys farreri, 49, 78
    Mimachlamys varia, 42, 44–50, 55, 56, 58, 59, 64, 66, 68, 70, 73, 74, 78, 84, 157–159, 166–168, 171, 173
    variegated, 58, 59
ship piloting, 96
sigma
  layers, 4, 60
  vertical levels, 4, 60
skimmer, 83
slop, 12
sludge, 29, 33
social media, 93
socio-economic, 22, 25, 29, 36
software, 2, 4, 5, 10, 15–17
spectrofluorometer, 44, 45
spreading, 29

stakeholders, 22, 30–32, 34, 37, 38, 89, 90, 93, 98–101
Station de Recherches Sous-marines et Océanographiques de Calvi (STARESO), 124
status
  chemical, 22, 24, 35, 36
  ecological, 24, 27, 30, 35, 36, 55, 56
  environmental, 103, 104
  good, 22, 24, 25, 27, 34–36
  health, 158, 161, 162
  water, 27, 35, 36
stockpile, 91, 99
storms, 80
strategy, 22, 33, 34
Stress on Stress (SoS), 106
superoxide dismutase (SOD), 44, 46, 47, 49, 107, 118
suprabenthic, 78

T

tabletop exercises, 100
Tagus, 6, 7, 10
  Estuary, 10, 91, 96
temperature, 24, 33, 149, 159, 161, 162, 167
terminal, 10, 11, 16
tests
  Bartlett, 45
  Kolmogorov–Smirnov, 45
  Levene's, 63
  Shapiro–Wilk normality, 63
  Student's t, 63
  Wilcoxon, 63
thiol group, 44
tide, 9, 10, 12, 14
  ebb, 79
  high, 58, 79
  intertidal areas, 60
  low, 58, 79
  macro-tidal, 80
  neap, 79
  slack, 79
  spring, 79, 80
  sub-tidal areas, 60
  tidal constituents, 60
Tier 1, 16, 100
Tier 2, 91
tourism, 21, 29, 30
trace elements, 58, 60, 63, 73, 104, 112, 113, 119–121
tracers, 8, 9, 11, 12, 55, 59, 60, 79, 80, 83
trade union, 32
training, 7, 9
transatlantic, 93
transport, 95, 98, 100
trophic
  chain, 145–147
  flow, 152
tug, 12

U, V

UV, 149
valve gape, 152
vanadium (V), 60, 61, 63, 65, 71, 73
variance, 45, 160, 171, 172
victim, 31

W

Warburg–Dickens–Horecker shunt, 78
waste, 56, 99
water
  -sediment, 73
  -sheds, 73
  bathing, 22, 23, 27, 29, 30, 35, 36, 38
  blackwater, 56
  brackish, 28
  commission, 25, 32, 33
  drinking, 23, 29, 33

flows, 59, 84
fluxes, 79
freshwater, 22, 24, 27, 38, 148
groundwater, 24, 34
inshore, 24
marine, 24, 26, 27, 35, 38
masses, 58, 79, 81
rainwater, 56–59, 65, 67, 69, 74, 76, 78, 79, 157, 159, 161–173
seawater, 145, 147–152
shellfish, 22, 23, 28, 35, 36
surface, 24, 29, 35
temperature, 75
territorial, 24, 30
transitional, 24
wastewater, 103–105, 115

waves, 4, 5, 14, 15
website, 91–93
wind, 9, 11, 14, 60, 75, 80, 81
speed, 9, 60
workshop, 15
wreck, 10

X, Y, Z

xenobiotic, 49, 107, 124
yachtsmen, 56
zinc (Zn), 60, 61, 63, 65, 67, 71, 73, 76, 77
zooplankton, 109, 110, 114, 152

Other titles from

iSTE

in

Ecological Science

2021

AMIARD Jean-Claude
Disarmament and Decommissioning in the Nuclear Domain
(Radioactive Risk SET – Volume 4)

JOLY Fernand, BOURRIÉ Guilhem
Mankind and Deserts 1: Deserts, Aridity, Exploration and Conquests
Mankind and Deserts 2: Water and Salts
Mankind and Deserts 3: Wind in Deserts and Civilizations

MUTTIN Frédéric, THOMAS Hélène
Marine Environmental Quality

2020

BRUSLÉ Jacques, QUIGNARD Jean-Pierre
Fish Behavior 1: Eco-ethology
Fish Behavior 2: Ethophysiology

LE FLOCH Stéphane
Remote Detection and Maritime Pollution: Chemical Spill Studies

MIGON Christophe, NIVAL Paul, SCIANDRA Antoine
The Mediterranean Sea in the Era of Global Change 1: 30 Years of Multidisciplinary Study of the Ligurian Sea
The Mediterranean Sea in the Era of Global Change 2: 30 Years of Multidisciplinary Study of the Ligurian Sea

ROSSIGNOL Jean-Yves
Climatic Impact of Activities: Methodological Guide for Analysis and Action

2019

AMIARD Jean-Claude
Industrial and Medical Nuclear Accidents: Environmental, Ecological, Health and Socio-economic Consequences (Radioactive Risk SET – Volume 2)
Nuclear Accidents: Prevention and Management of an Accidental Crisis (Radioactive Risk SET – Volume 3)

BOULEAU Gabrielle
Politicization of Ecological Issues: From Environmental Forms to Environmental Motives

DAVID Valérie
Statistics in Environmental Sciences

GIRAULT Yves
UNESCO Global Geoparks: Tension Between Territorial Development and Heritage Enhancement

KARA Mohamed Hichem, QUIGNARD Jean-Pierre
Fishes in Lagoons and Estuaries in the Mediterranean 2: Sedentary Fish
Fishes in Lagoons and Estuaries in the Mediterranean 3A: Migratory Fish
Fishes in Lagoons and Estuaries in the Mediterranean 3B: Migratory Fish

OUVRARD Benjamin, STENGER Anne
Incentives and Environmental Policies: From Theory to Empirical Novelties

2018

AMIARD Jean-Claude
Military Nuclear Accidents: Environmental, Ecological, Health and Socio-economic Consequences
(Radioactive Risk SET – Volume 1)

FLIPO Fabrice
The Coming Authoritarian Ecology

GUILLOUX Bleuenn
Marine Genetic Resources, R&D and the Law 1: Complex Objects of Use

KARA Mohamed Hichem, QUIGNARD Jean-Pierre
Fishes in Lagoons and Estuaries in the Mediterranean 1: Diversity, Bioecology and Exploitation

2016

BAGNÈRES Anne-Geneviève, HOSSAERT-MCKEY Martine
Chemical Ecology

2014

DE LARMINAT Philippe
Climate Change: Identification and Projections

Printed and bound by CPI Group (UK) Ltd, Croydon, CR0 4YY

15/09/2026

14962407-0003